E mail us at **design.te0m@gmail.com** for any comments about the book which will allow us to evaluate our work or leave us a comment.

Let's go to training your child's hand

THIS BOOK
BELONGS TO

1  2  3

# Table of Contents

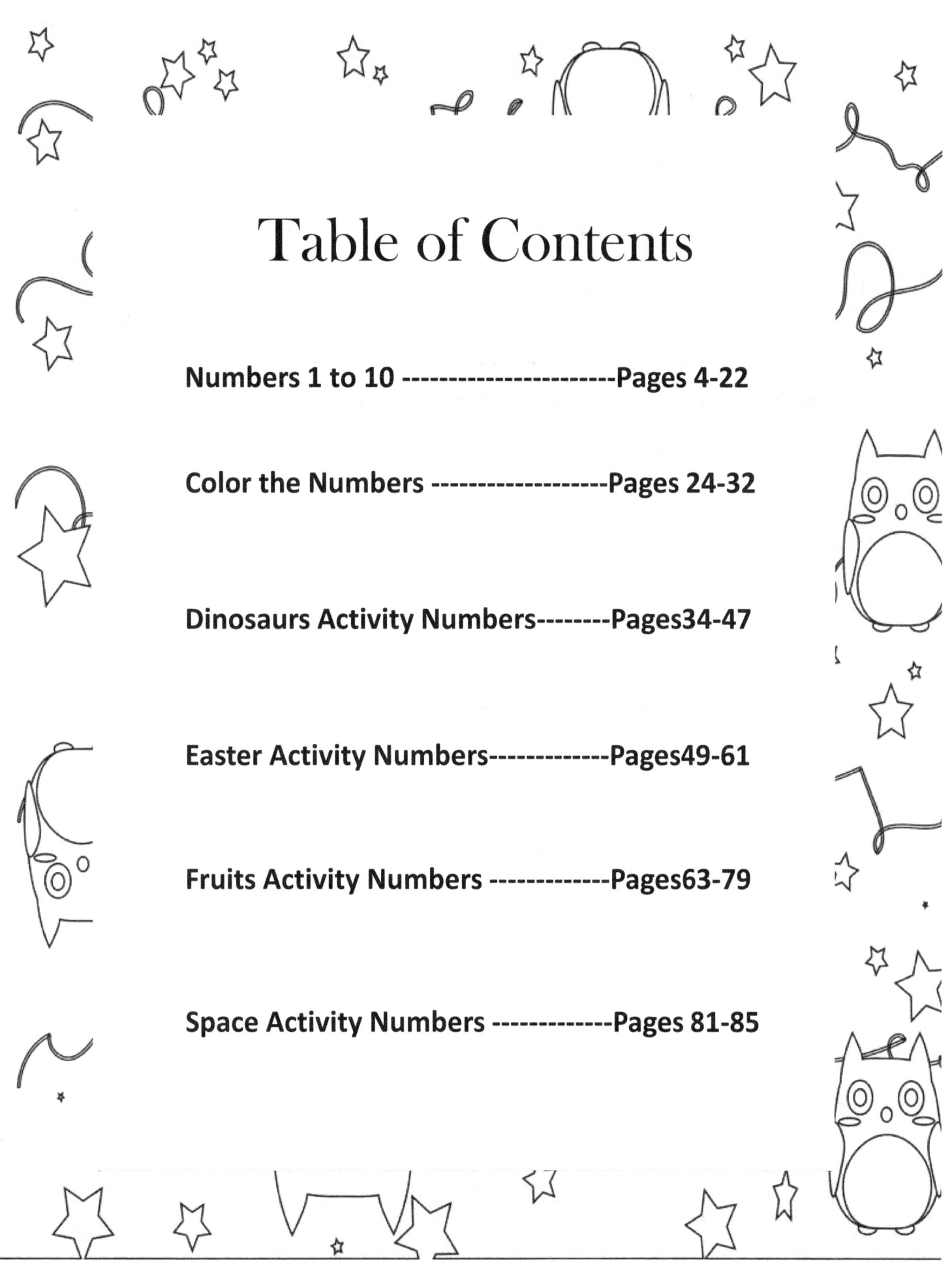

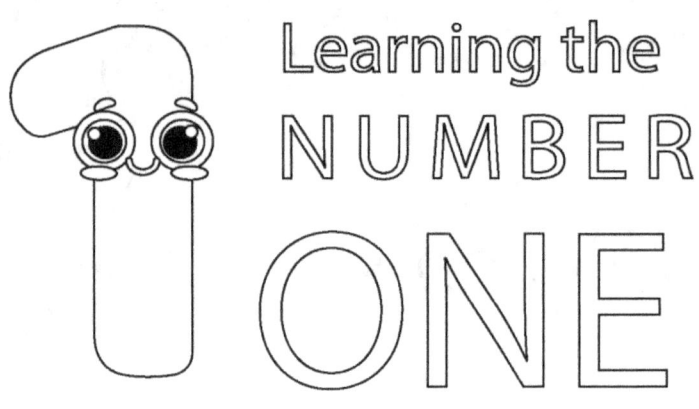

# Learning the
# NUMBER
# ONE

**There is ONE whale**

---

**Mark the right amount showed in the hands:**

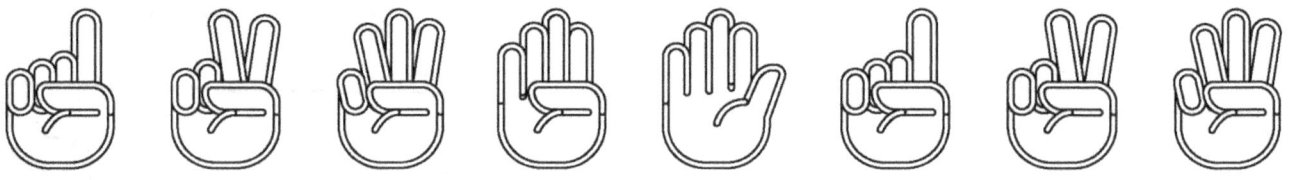

---

**Trace the number 1:**

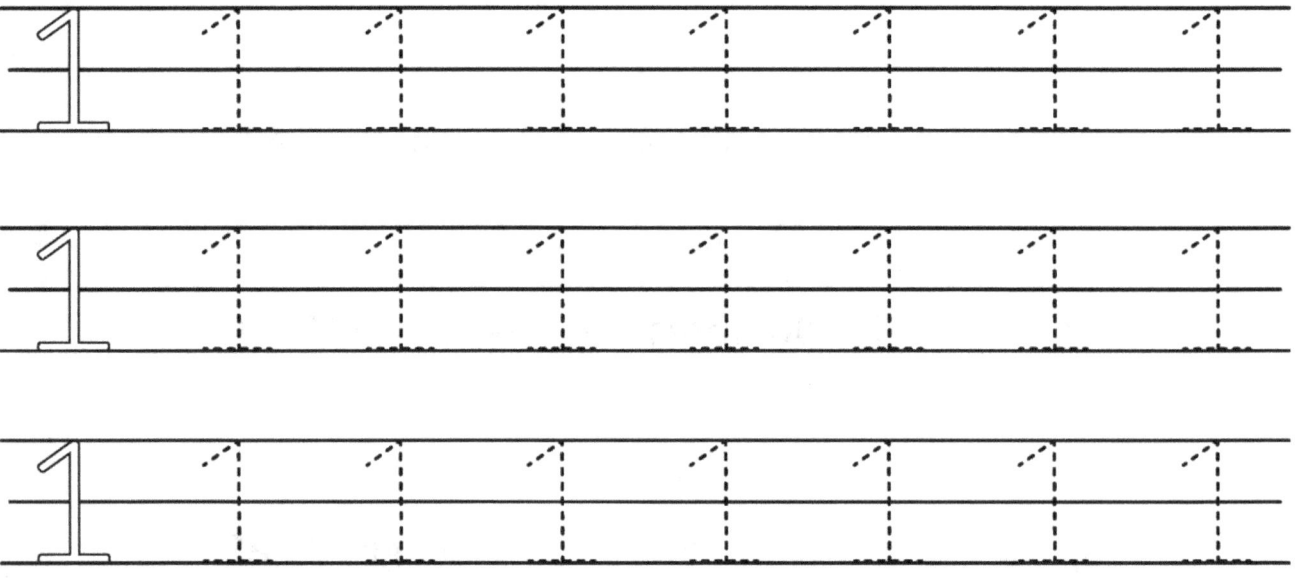

---

**Color ONE star:**

# Worksheet

**Find and color 2:**

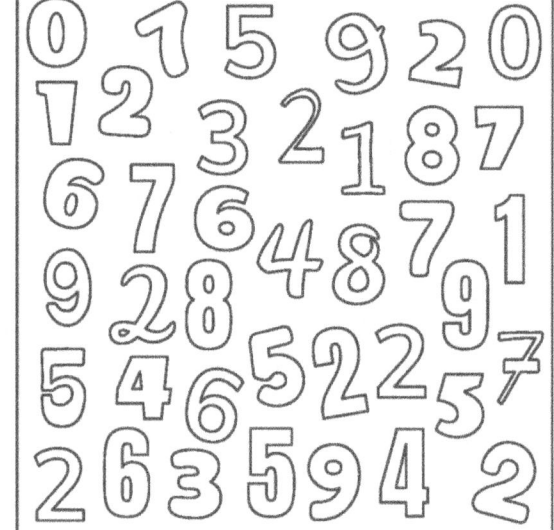

2 2 2 2 2 2 2 2 2 2 2

2

# LET'S LEARN THE NUMBER
# THREE

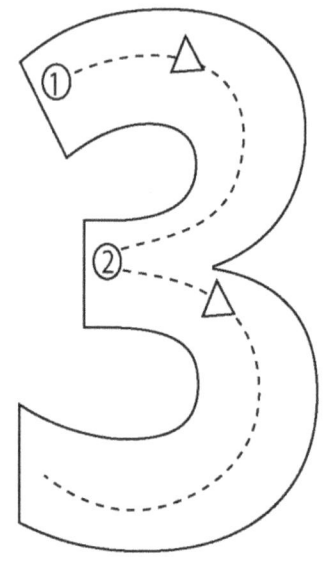

## TRACE THE NUMBER 3

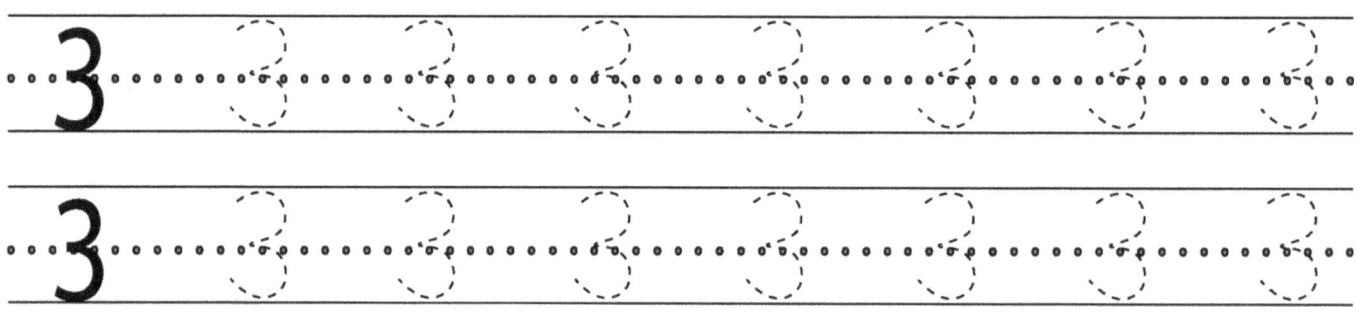

## COLOR THREE

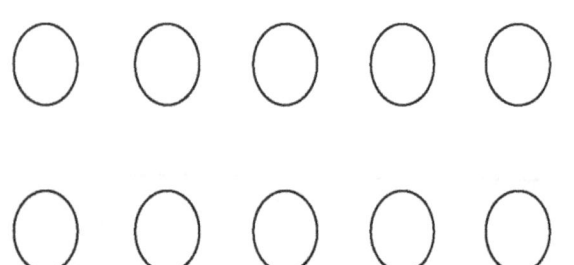

## CIRCLE THE WORD "THREE"

# LET'S LEARN THE NUMBER
# FOUR

## TRACE THE NUMBER 4

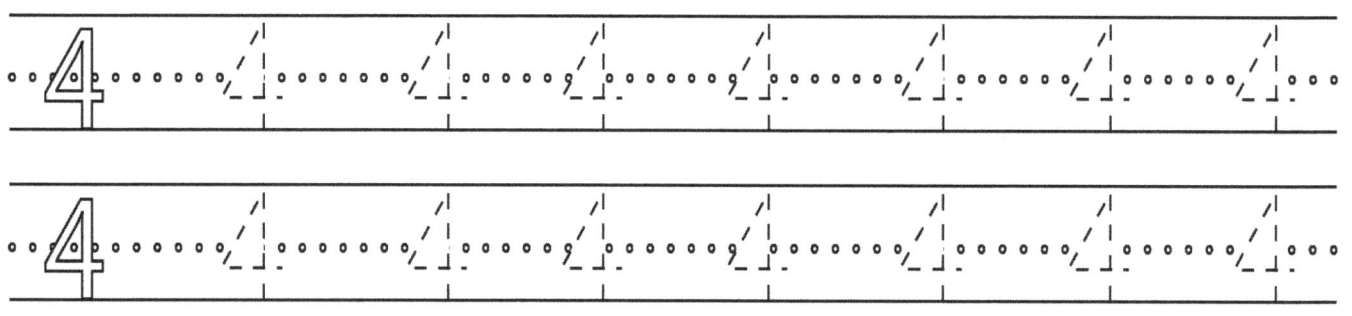

## COLOR FOUR

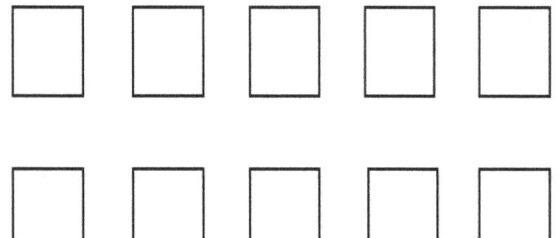

## CIRCLE THE NUMBER 4's

## Let's learn the number FIVE

Trace the number

5 5 5 5 5 5 5 5

5 5 5 5 5 5 5 5

Color 5 hearts

How many dots are in the box?

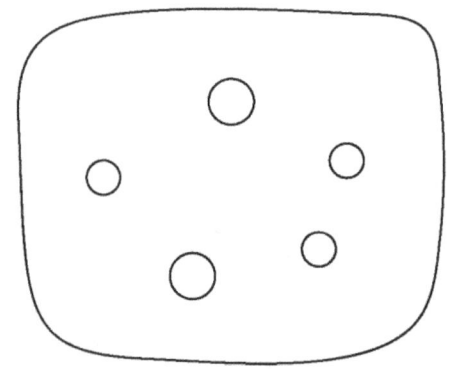

# LET'S LEARN THE NUMBER

## SIX

## TRACE THE NUMBER 6

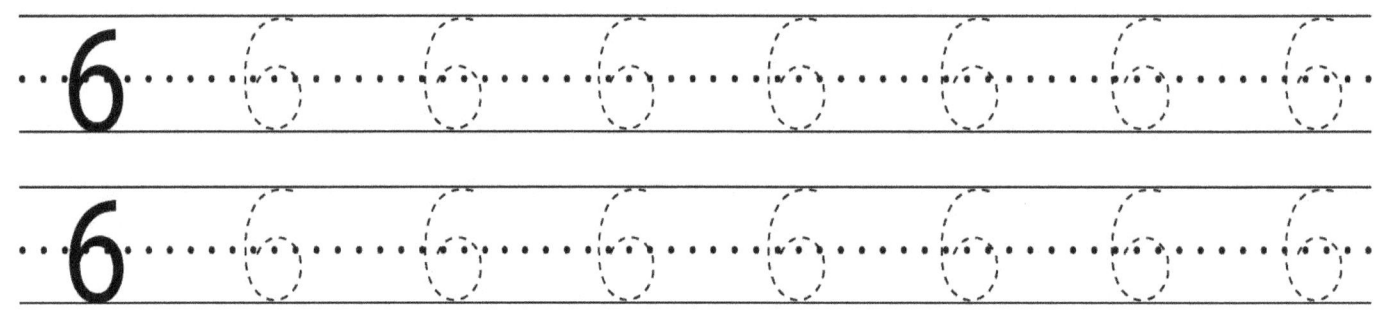

## COLOR SIX

## CIRCLE THE WORD "SIX"

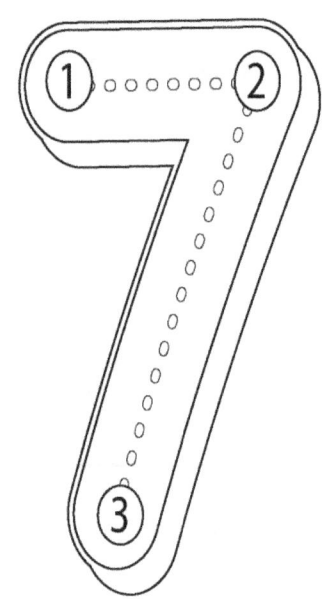

# LET'S LEARN
## THE NUMBER SEVEN

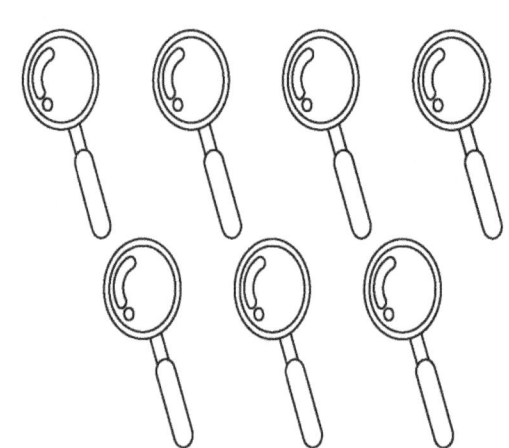

## TRACE THE NUMBER 7

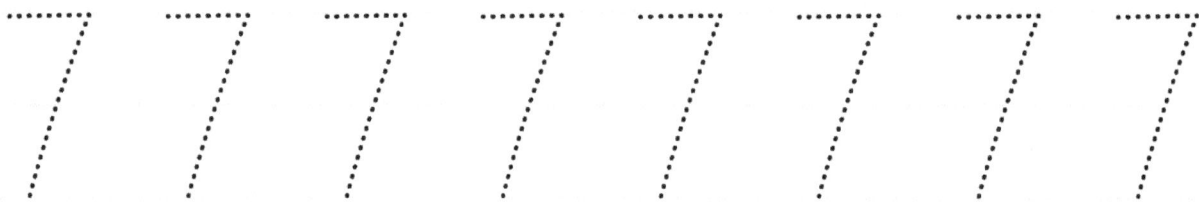

## COLOR 7 CIRCLES

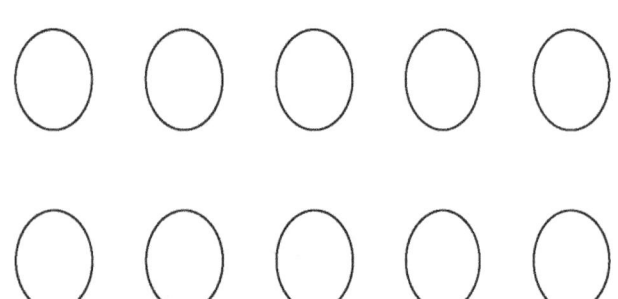

## CIRCLE THE NUMBER 7

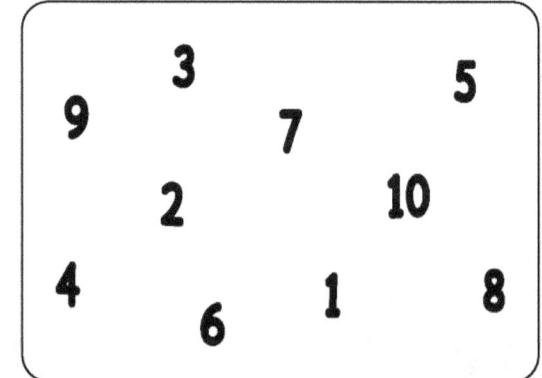

# LET'S LEARN THE NUMBER

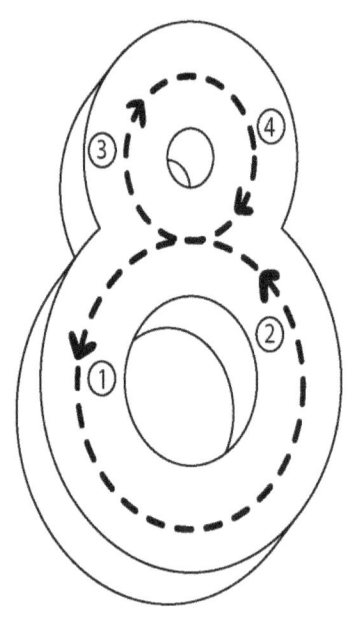

## LET'S TRACE IT!

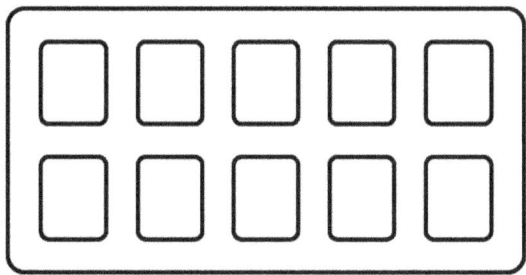

## CIRCLE THE NUMBER 8's

1 2 3 4 5
6 7 8 9 10

## COLOR 8 OBJECTS

## CIRCLE THE BOX WITH 8 IMAGES

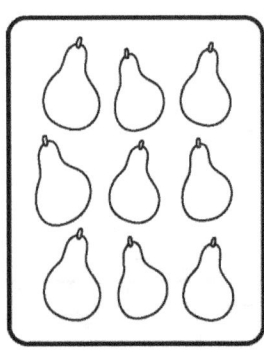
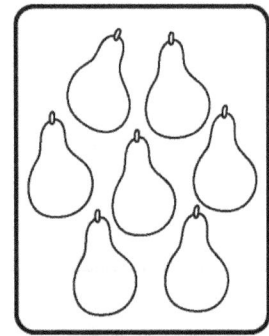
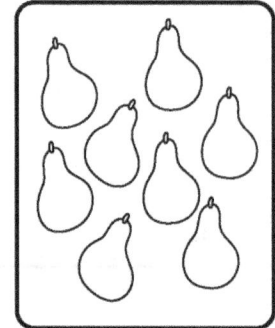
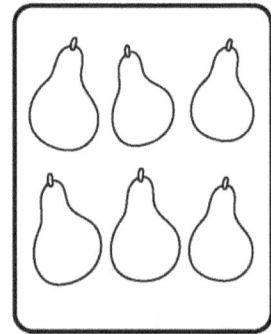

18

## Let's learn the number NINE

### Trace the number

9 9 9 9 9 9 9 9 9
9 9 9 9 9 9 9 9 9

### Color 9 clouds

### Find the word "NINE"

| F | I | V | E |
|---|---|---|---|
| N | I | N | E |
| F | O | U | R |

TEN

## Color the starfishes with number TEN

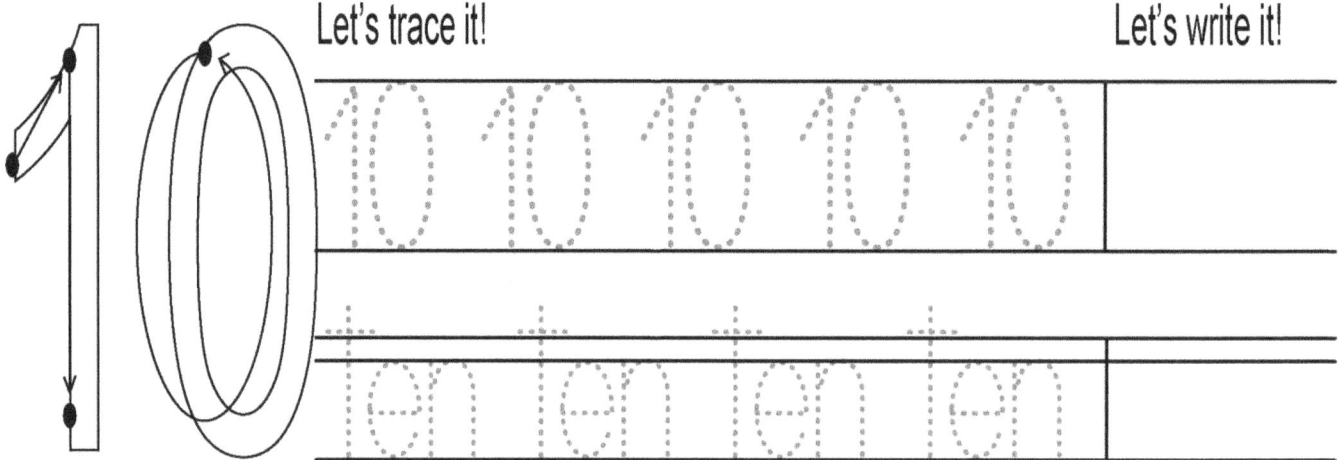

**Let's trace it!**

10 10 10 10 10

ten ten ten ten

**Let's write it!**

## Circle the box with TEN starfish

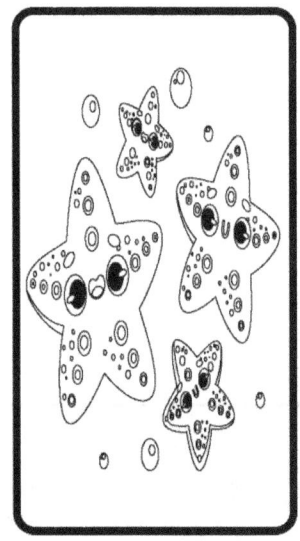
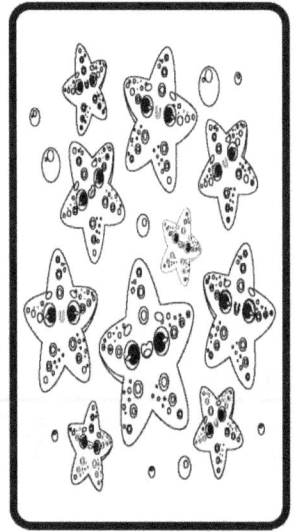
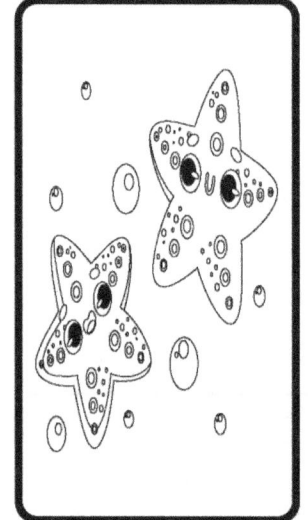
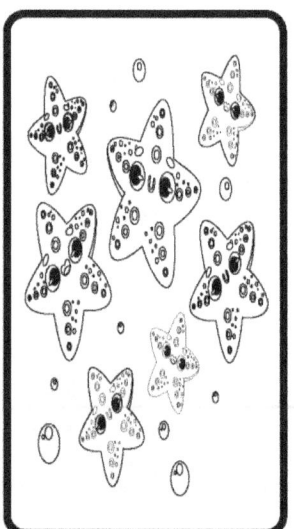

22

Color the number of beads in the abacus to equal the number
that is in the box on the right side of the abacus.

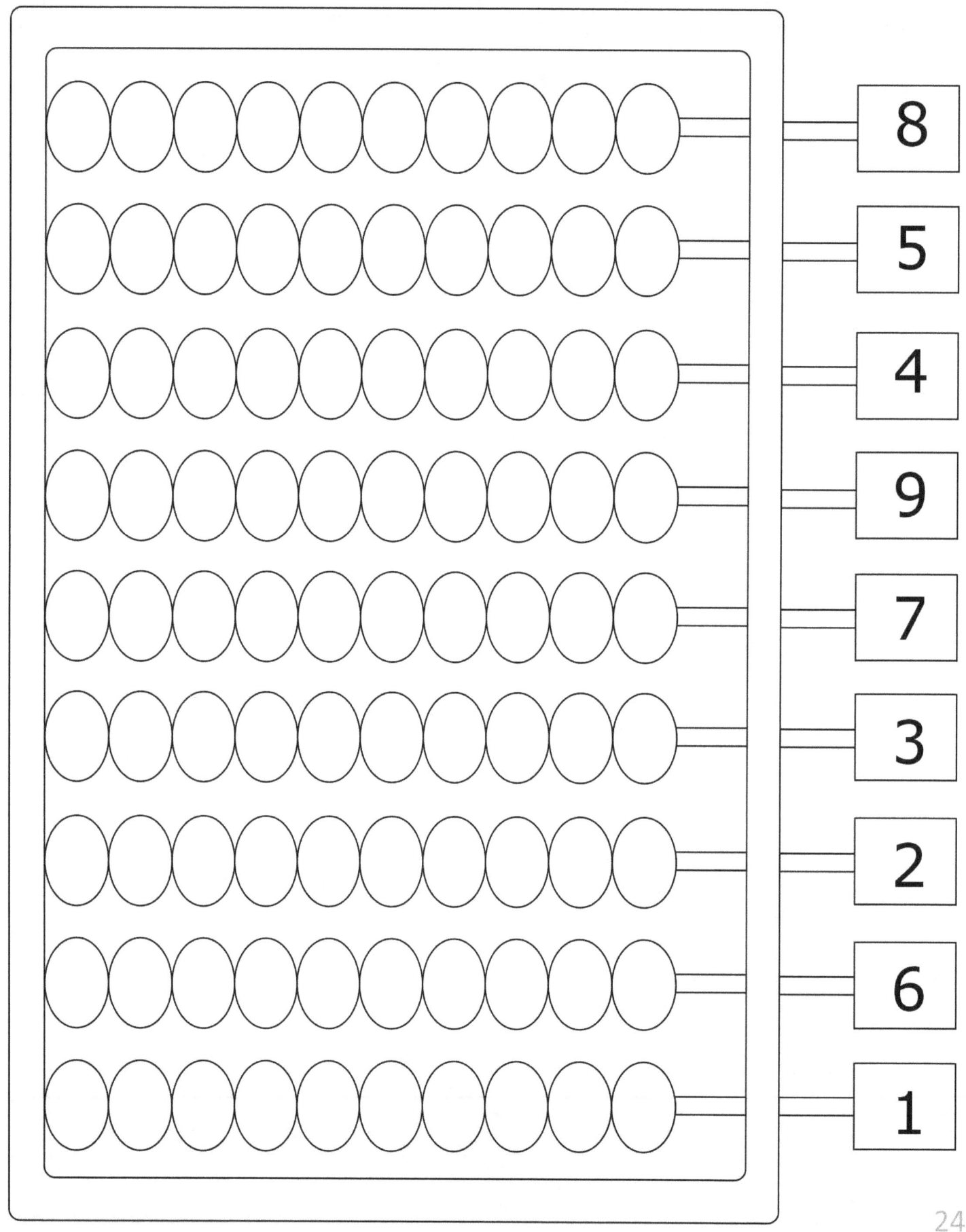

Color the number of beads in the abacus to equal the number
that is in the box on the right side of the abacus.

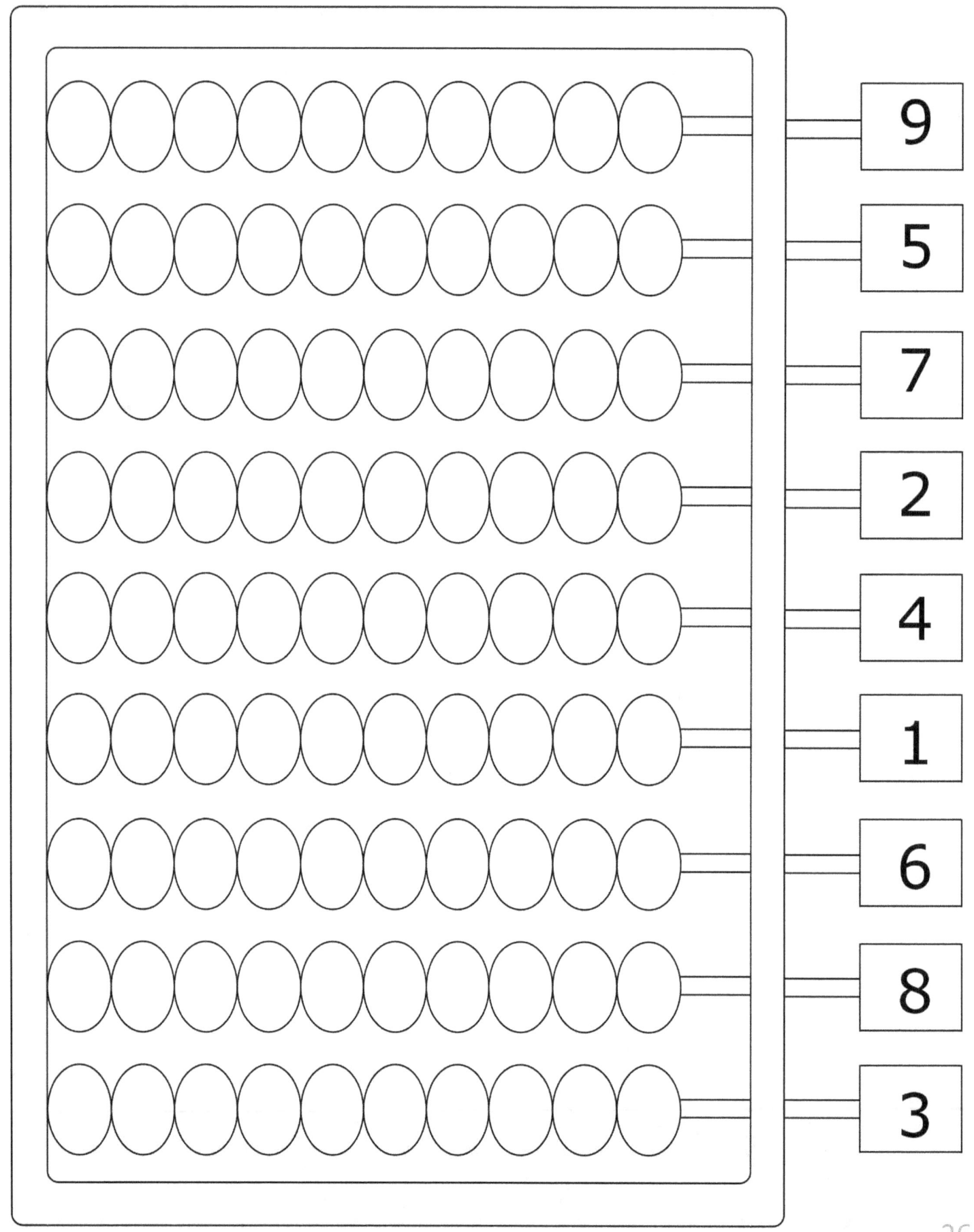

Color the number of beads in the abacus to equal the number
that is in the box on the right side of the abacus.

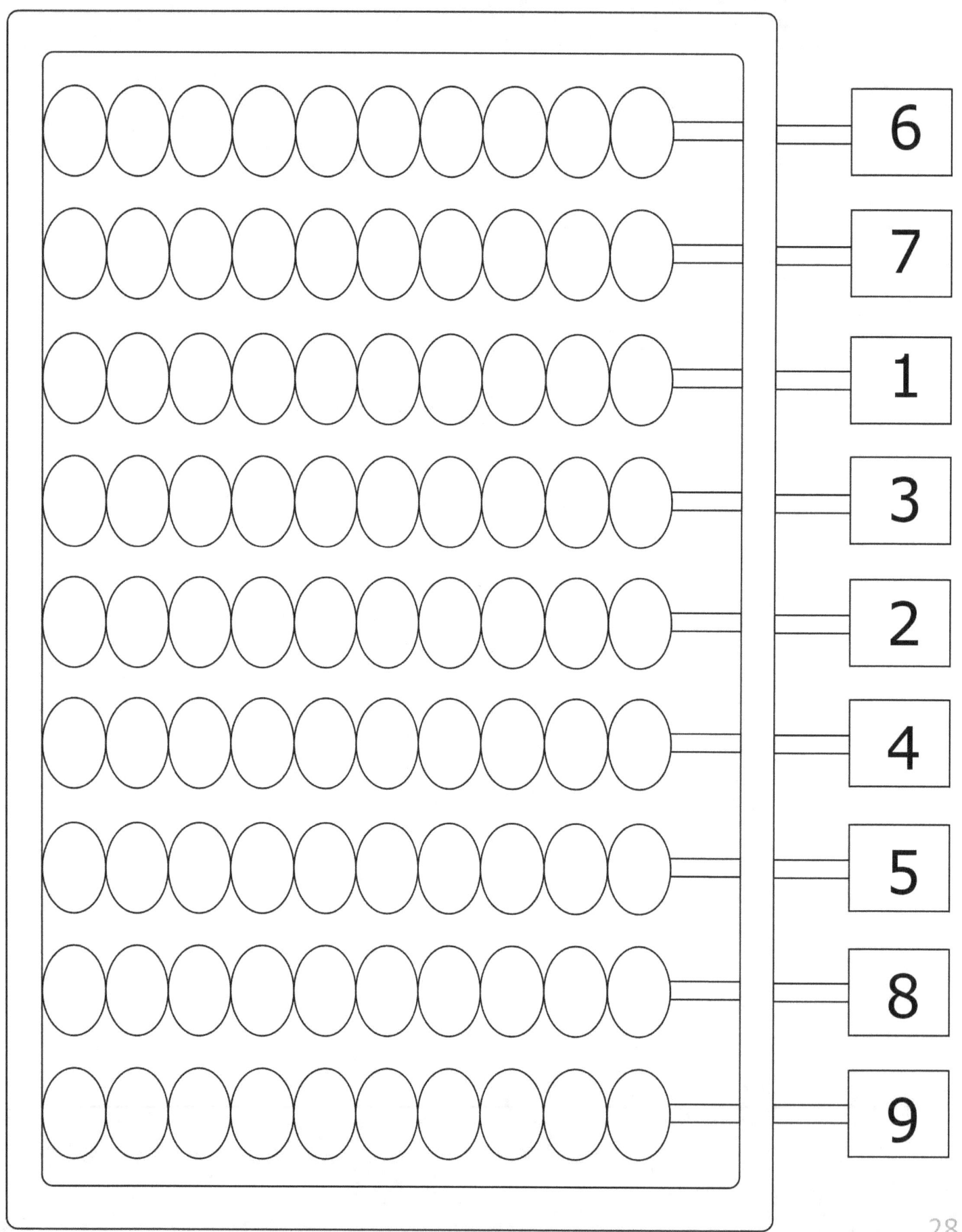

Color the number of beads in the abacus to equal the number
that is in the box on the right side of the abacus.

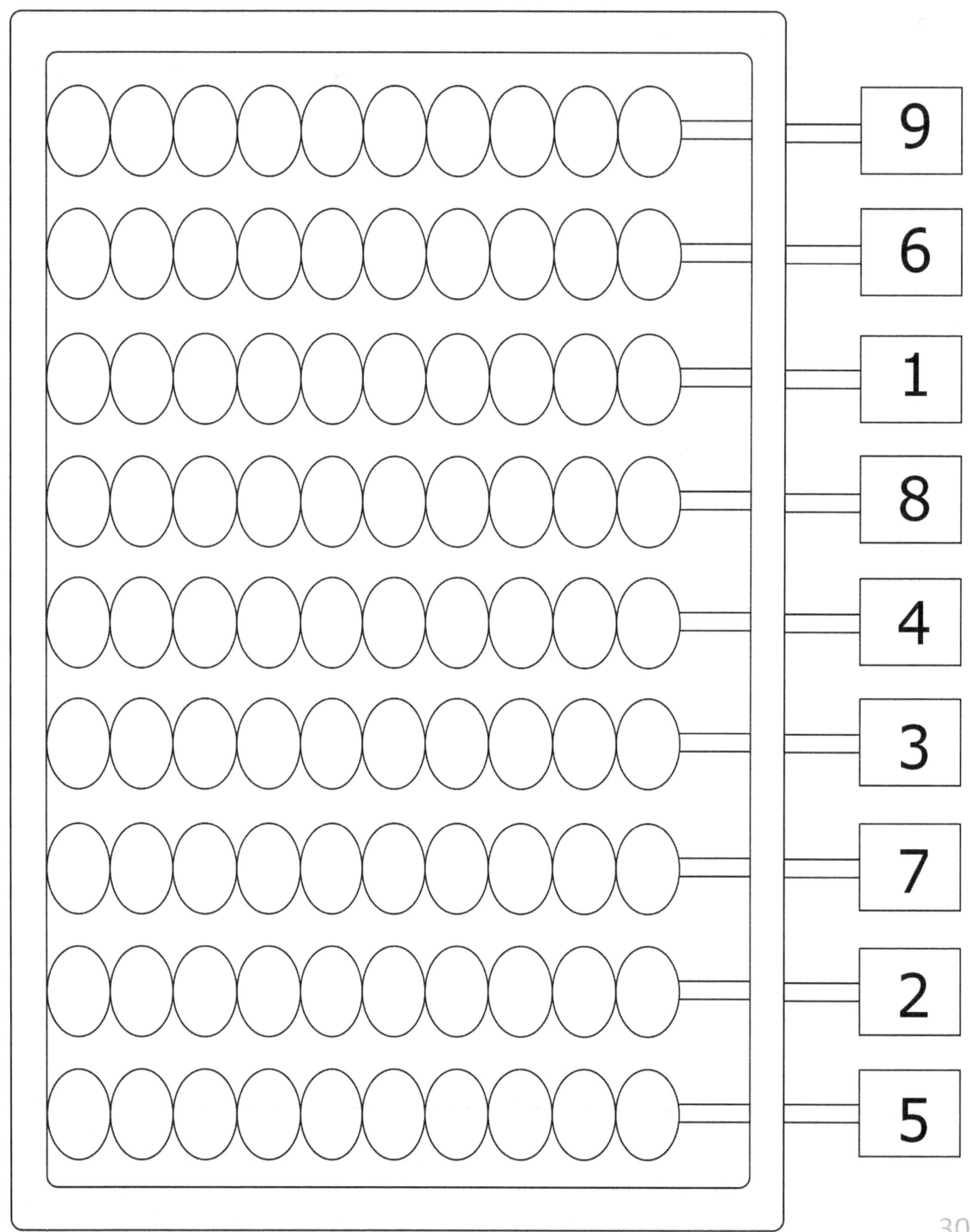

Color the number of beads in the abacus to equal the number
that is in the box on the right side of the abacus.

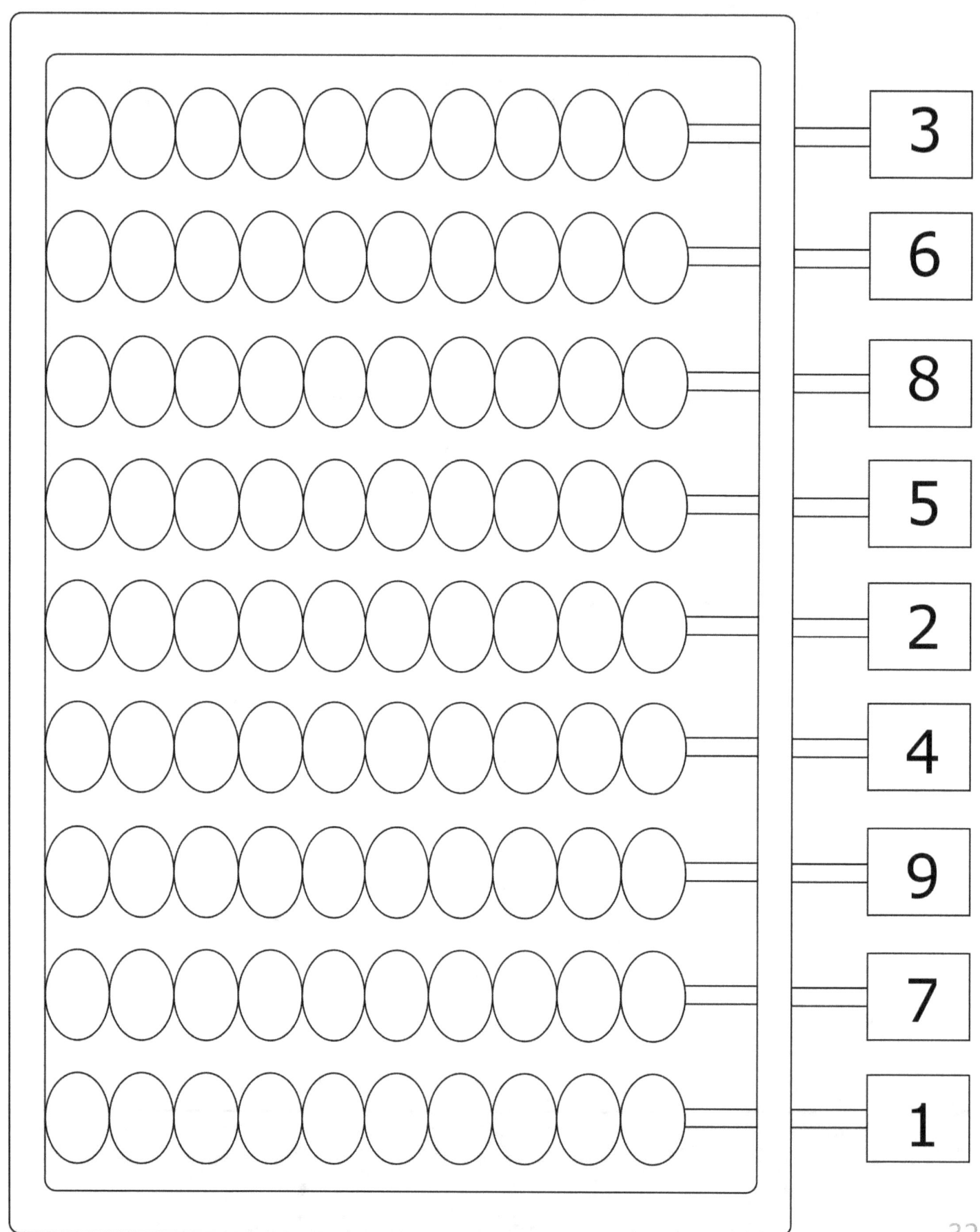

# Connects the number to the image

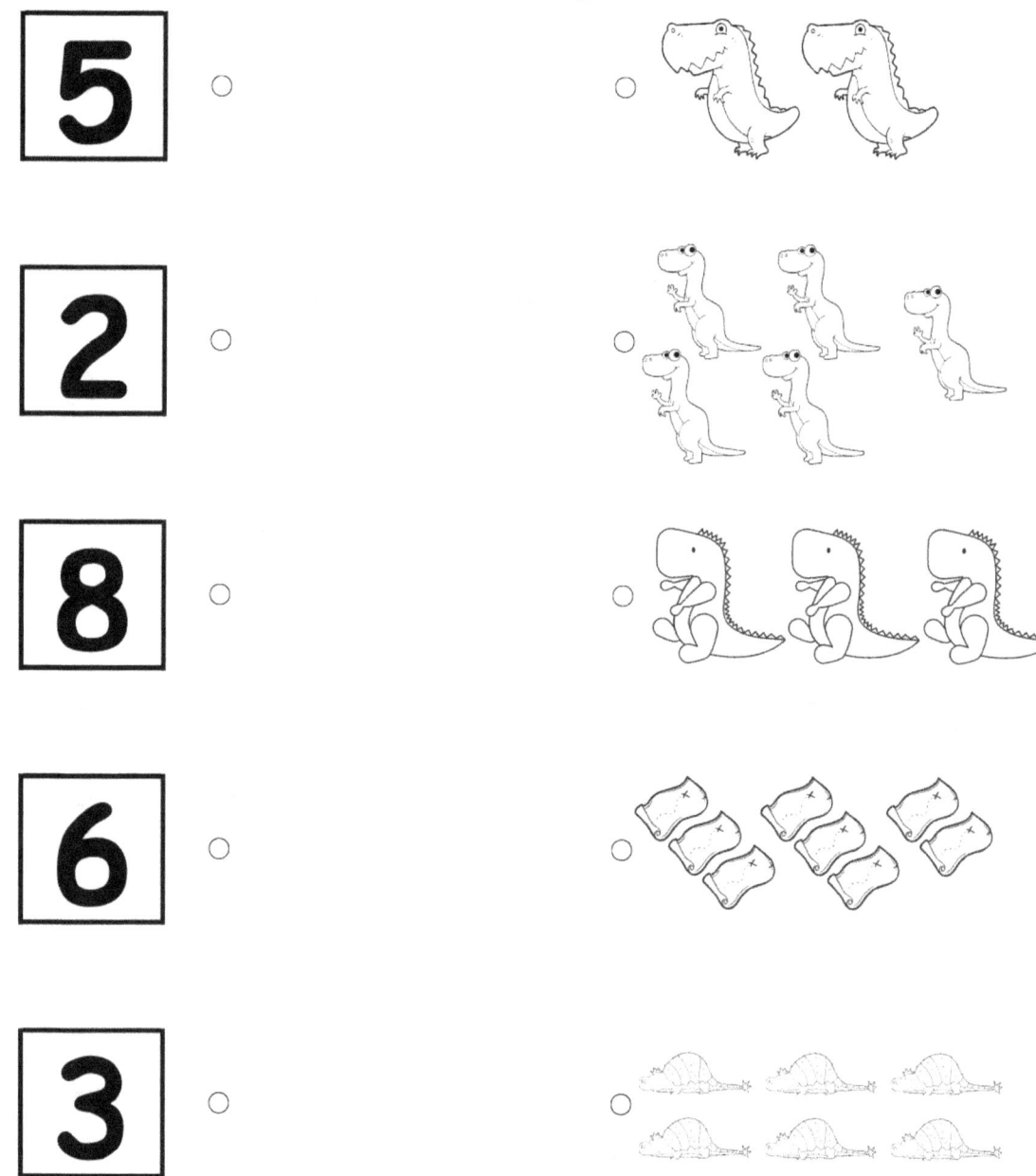

5

2

8

6

3

# Connects the number to the image

1 ○                              ○

2 ○                              ○

3 ○                              ○

4 ○                              ○

5 ○                              ○

# Connects the number to the image

1

5

4

2

7

# Connects the number to the image

2 ○  ○

8 ○  ○

5 ○  ○

7 ○  ○

1 ○  ○

# Connects the number to the image

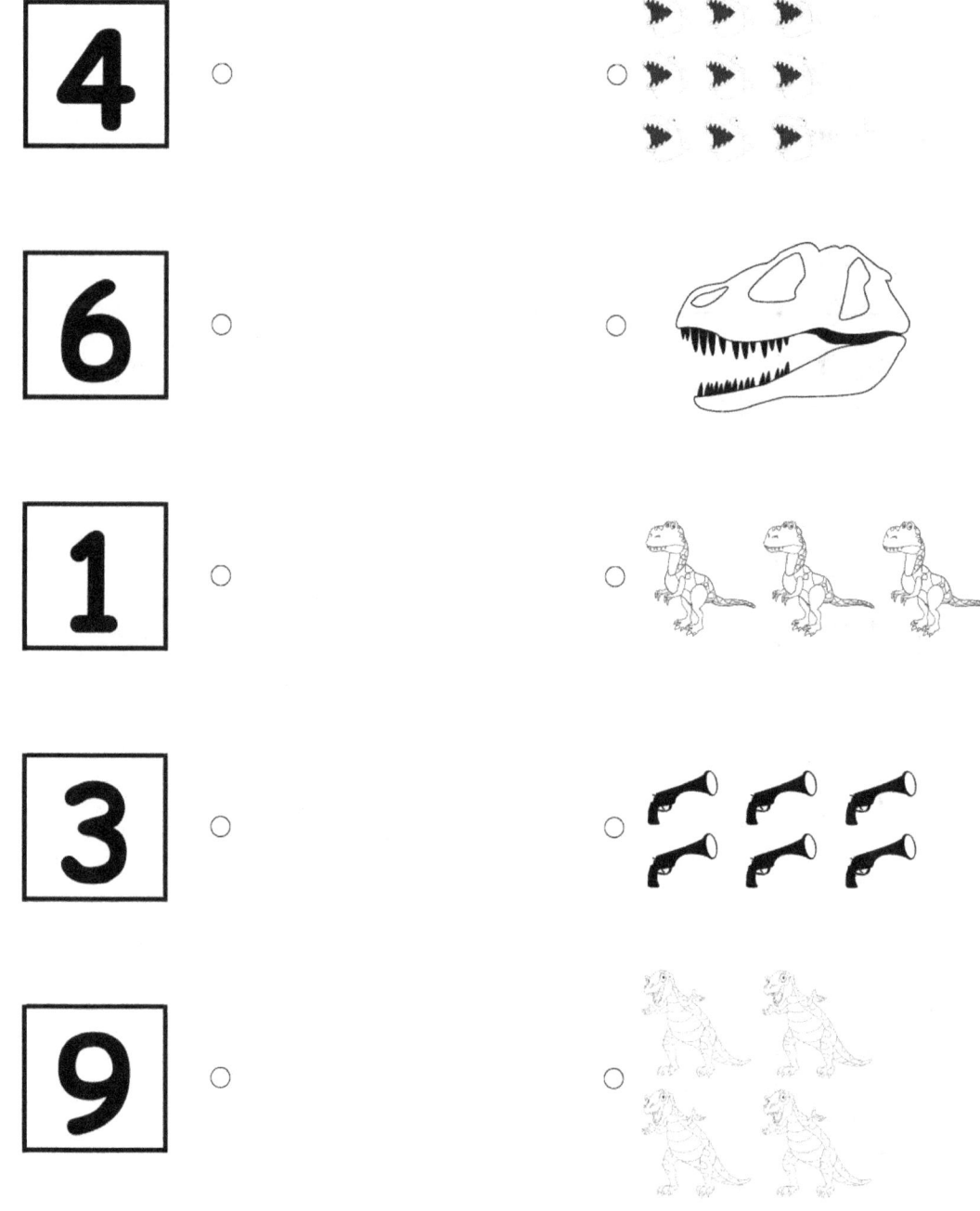

4

6

1

3

9

# Magical colorings

6 + 3 = ☐ yellow    1 + 2 = ☐ blue

2 + 5 = ☐ green    1 + 4 = ☐ orange

3 + 5 = ☐ red    2 + 2 = ☐ pink

39

# Magical colorings

6 + 3 = [ ]  yellow       1 + 2 = [ ]  blue

2 + 5 = [ ]  green        1 + 4 = [ ]  orange

3 + 5 = [ ]  red          2 + 2 = [ ]  pink

# Magical colorings

6 + 3 = [ ] yellow     1 + 2 = [ ] blue

2 + 5 = [ ] green     1 + 4 = [ ] orange

3 + 5 = [ ] red     2 + 2 = [ ] pink

# Magical colorings

6 + 3 = [ ] yellow    1 + 2 = [ ] blue

2 + 5 = [ ] green    1 + 4 = [ ] orange

3 + 5 = [ ] red    2 + 2 = [ ] pink

# Magical colorings

6 + 3 = [  ] yellow    1 + 2 = [  ] blue

2 + 5 = [  ] green    1 + 4 = [  ] orange

3 + 5 = [  ] red    2 + 2 = [  ] pink

# EASTER COUNT THE NUMBERS

Count the images and write how many in the box

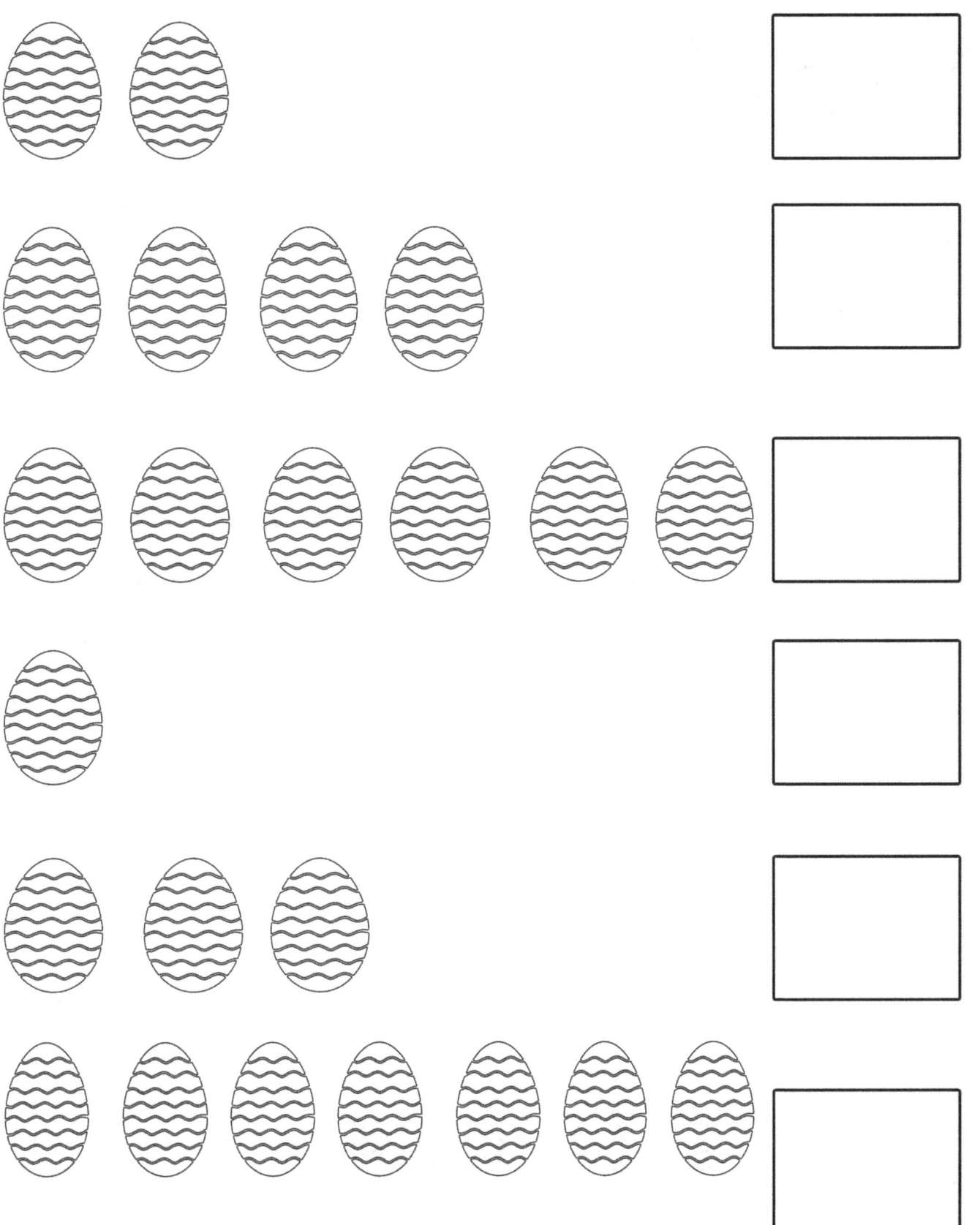

# EASTER COUNT THE NUMBERS

Count the images and write how many in the box

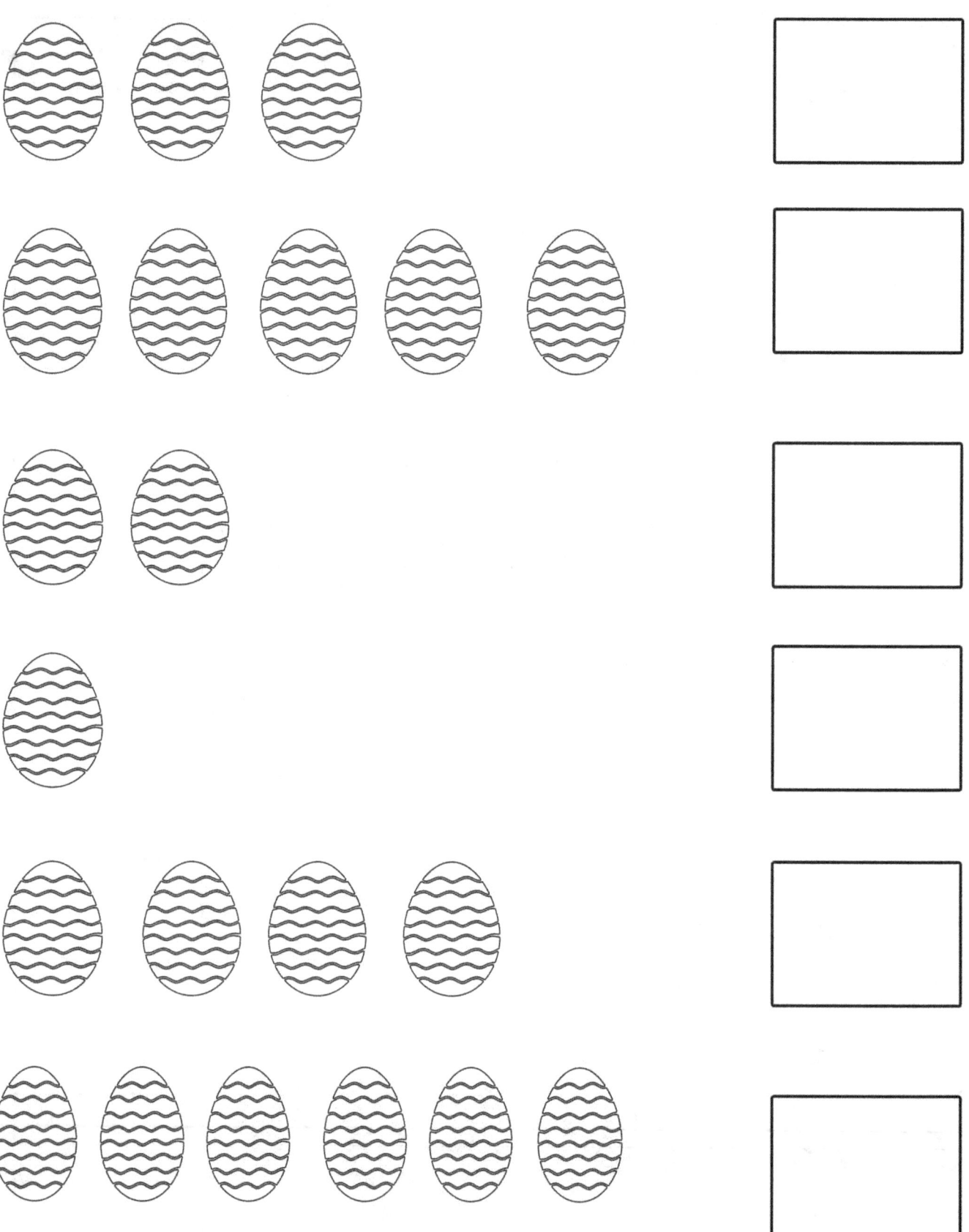

# EASTER COUNT THE NUMBERS

Count the images and write how many in the box

# EASTER COUNT THE NUMBERS

Count the images and write how many in the box

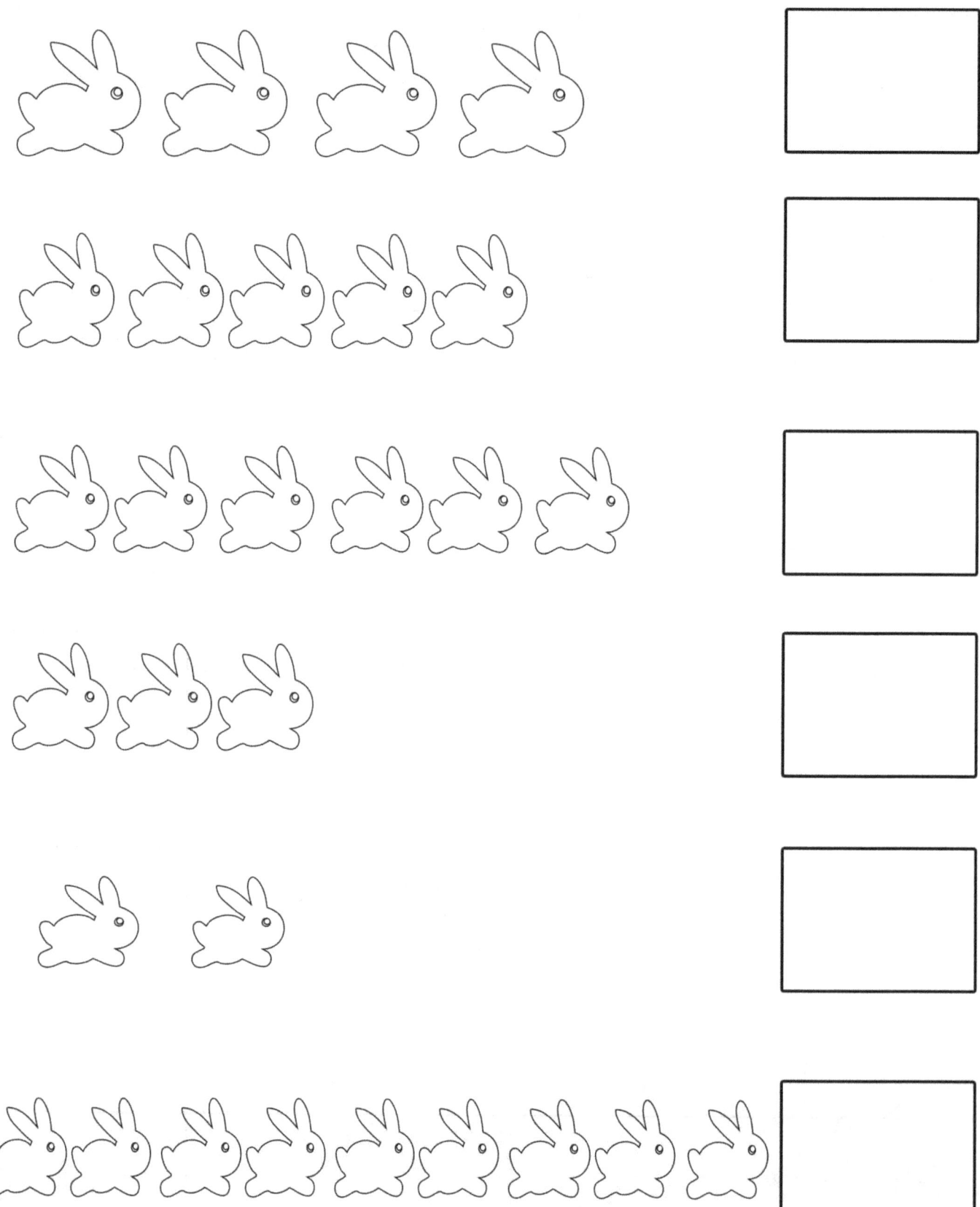

# EASTER COUNT THE NUMBERS

CIRCLE THE NUMBER TO SHOW HOW MANY

4 6 8 2

2 3 7 9

5 9 8 1

6 2 3 1

3 6 8 5

6 7 8 5

# EASTER COUNT THE NUMBERS

CIRCLE THE NUMBER TO SHOW HOW MANY

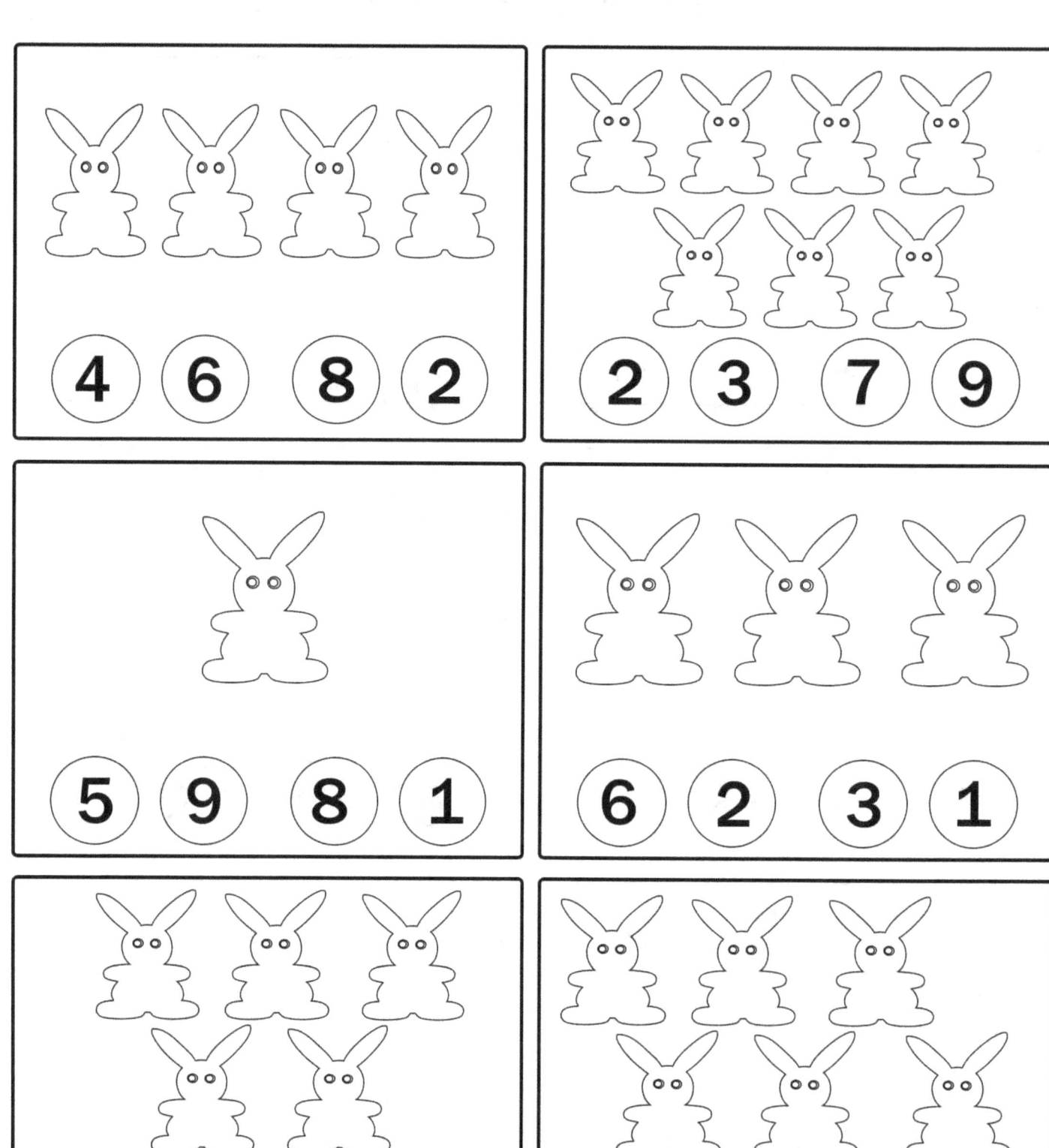

# EASTER COUNT THE NUMBERS

COLOR THE NUMBER TO SHOW HOW MANY

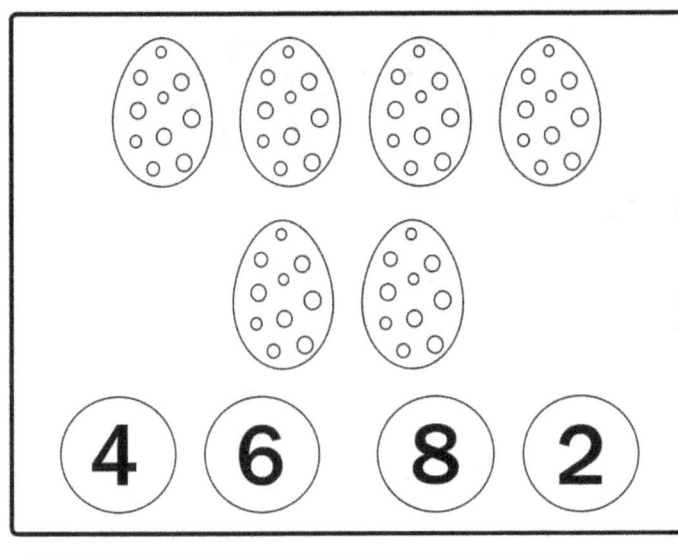

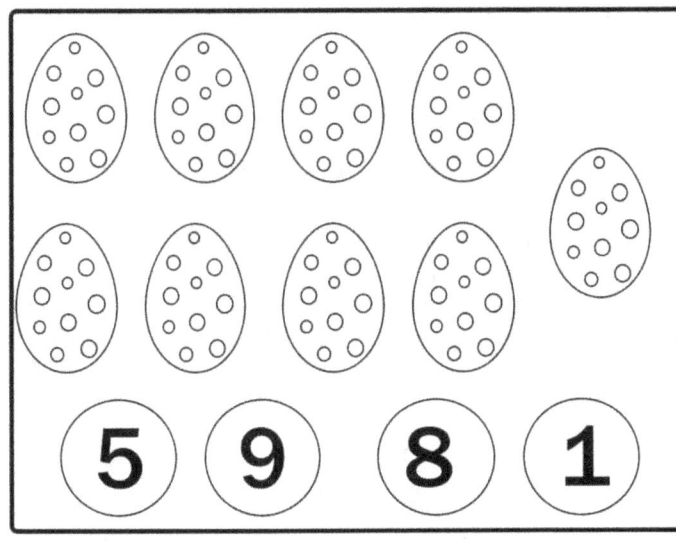

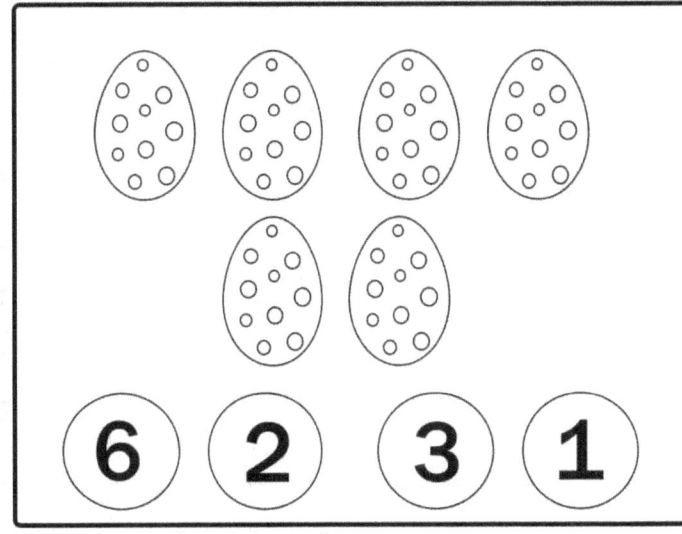

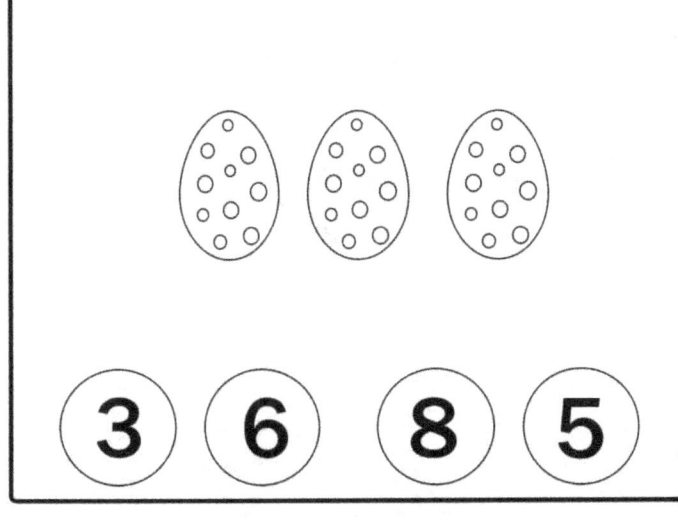

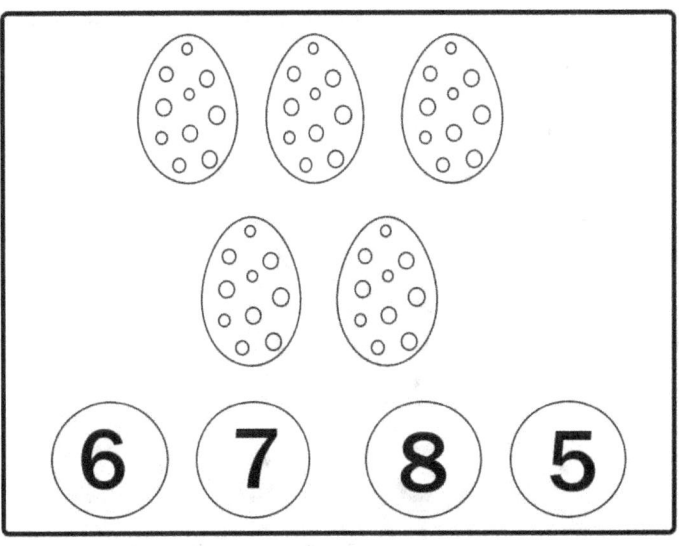

# EASTER COUNT THE NUMBERS

COLOR THE NUMBER TO SHOW HOW MANY

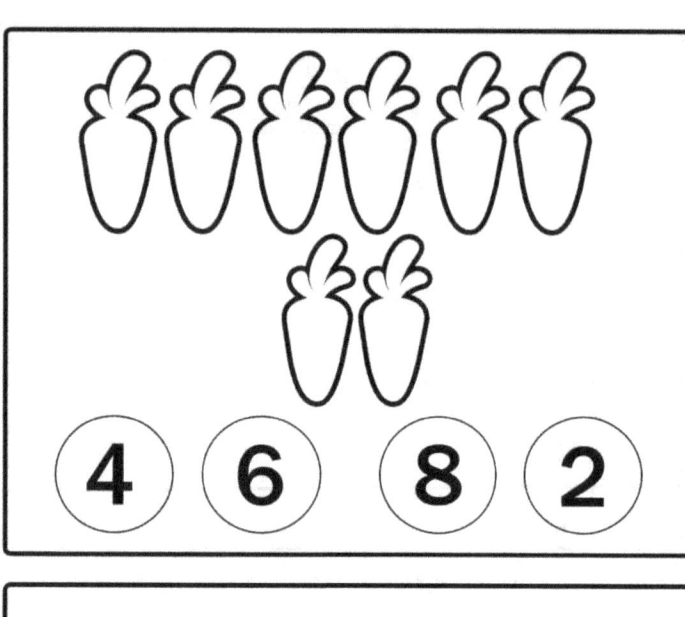

4   6   8   2

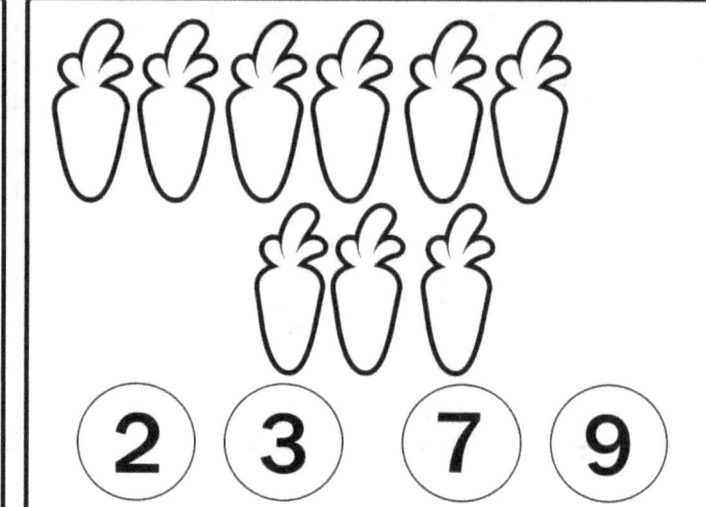

2   3   7   9

5   9   8   1

6   2   3   1

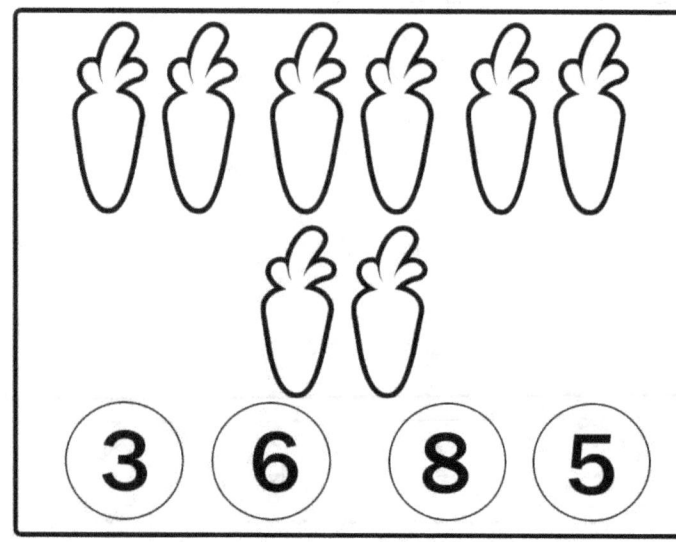

3   6   8   5

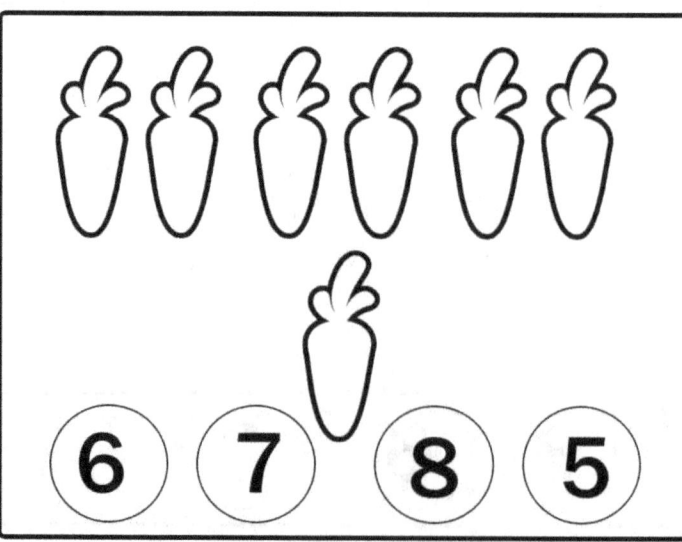

6   7   8   5

56

# EASTER COUNT THE NUMBERS

FILL IN THE MISSING NUMBERS

# EASTER COUNT THE NUMBERS

FILL IN THE MISSING NUMBERS

# COLOR THE FRAMES TO MATCH THE NUMBERS ON THE PIC

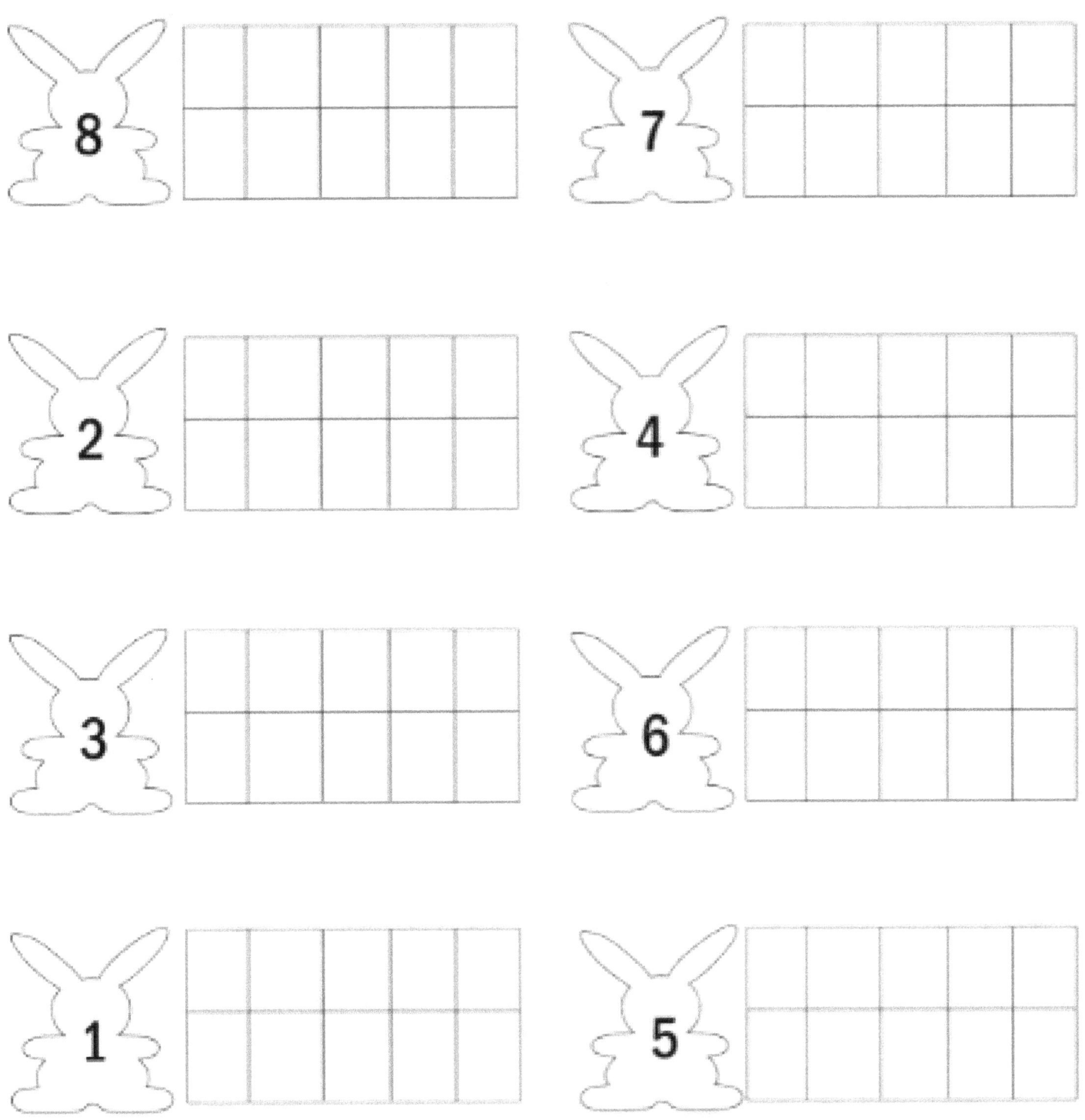

# COLOR THE FRAMES TO MATCH THE NUMBERS ON THE PIC

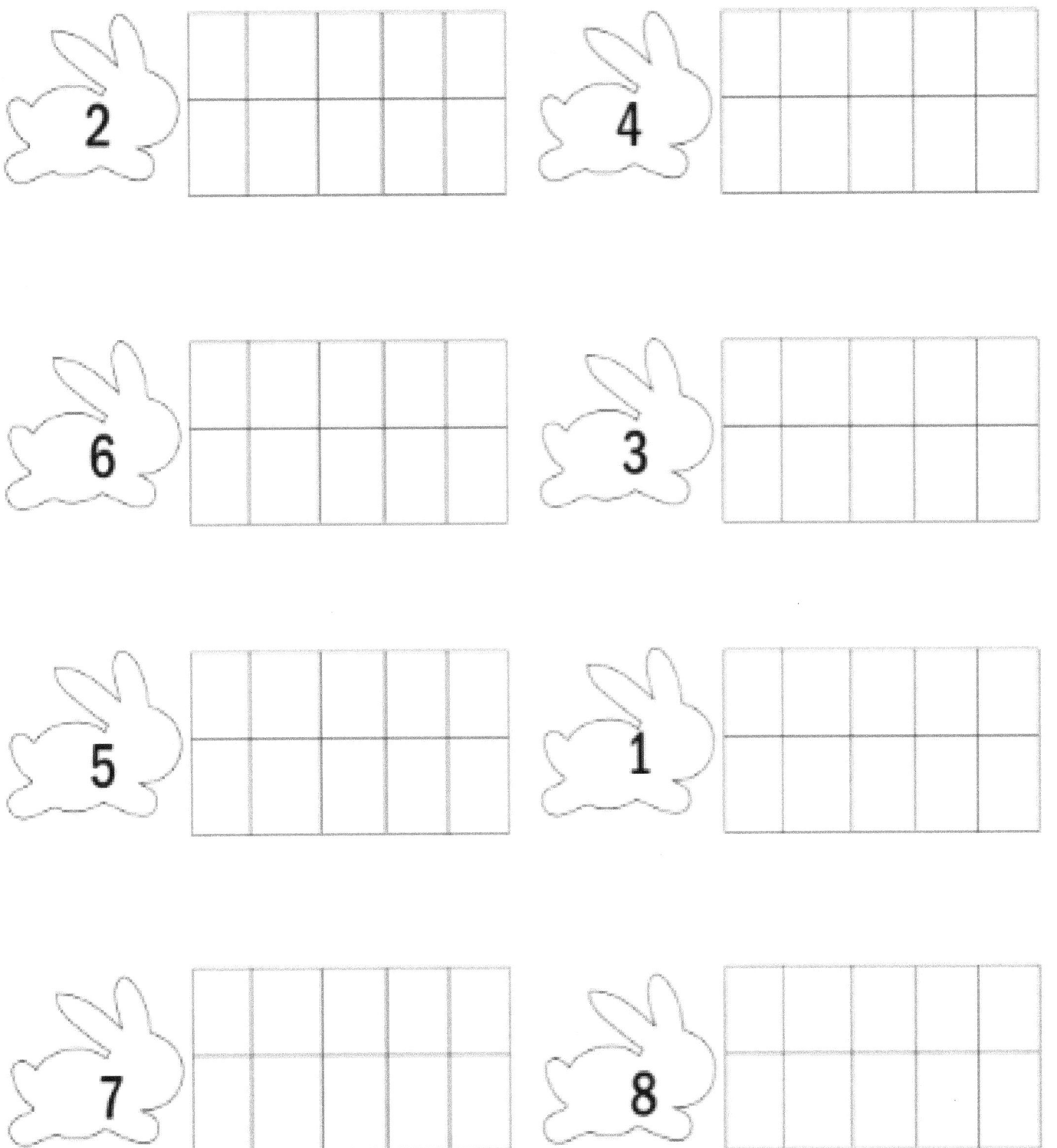

# COLOR THE FRAMES TO MATCH THE NUMBERS ON THE FRUIT

1

5

9

6

2

4

7

8

3

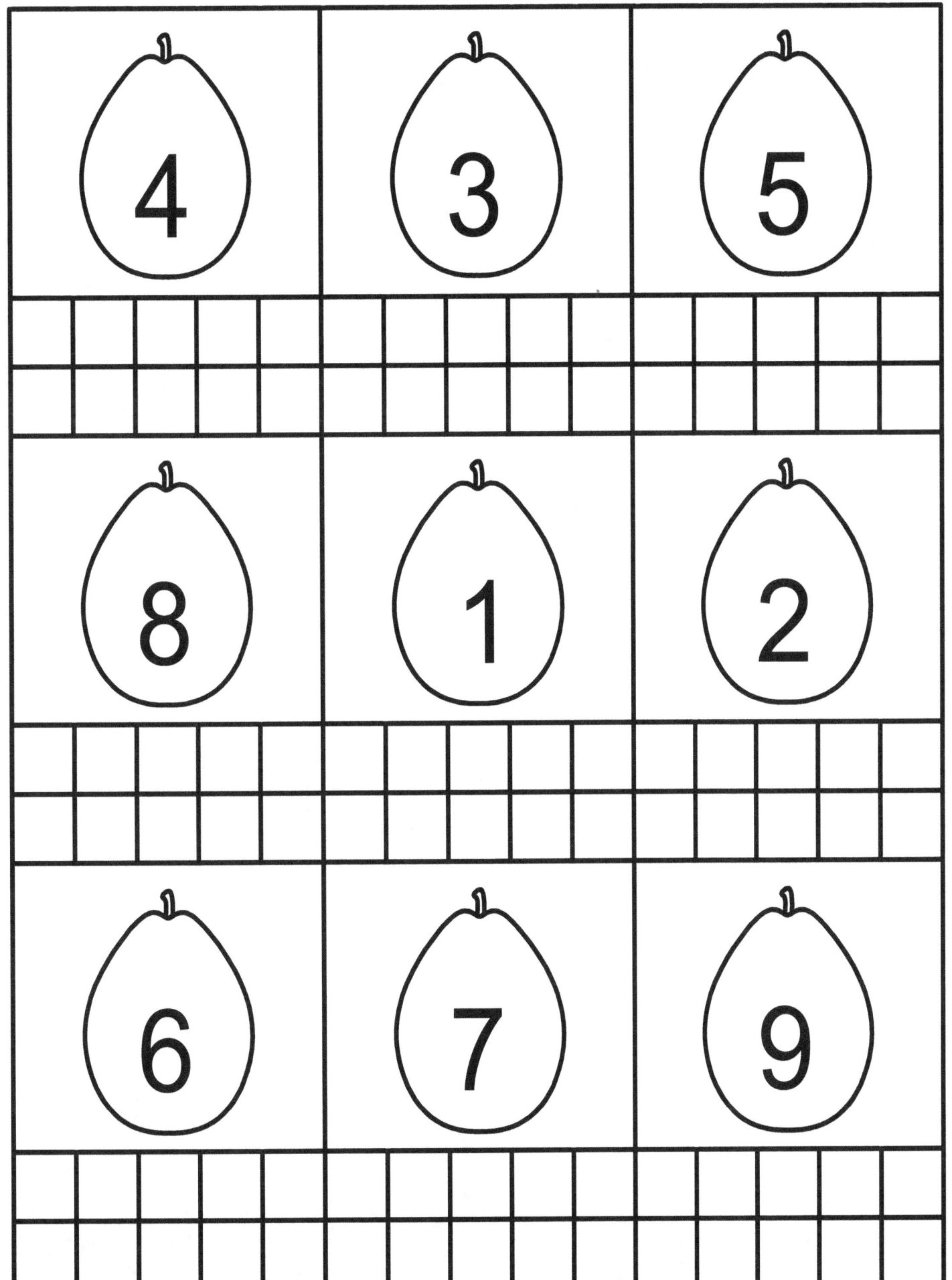

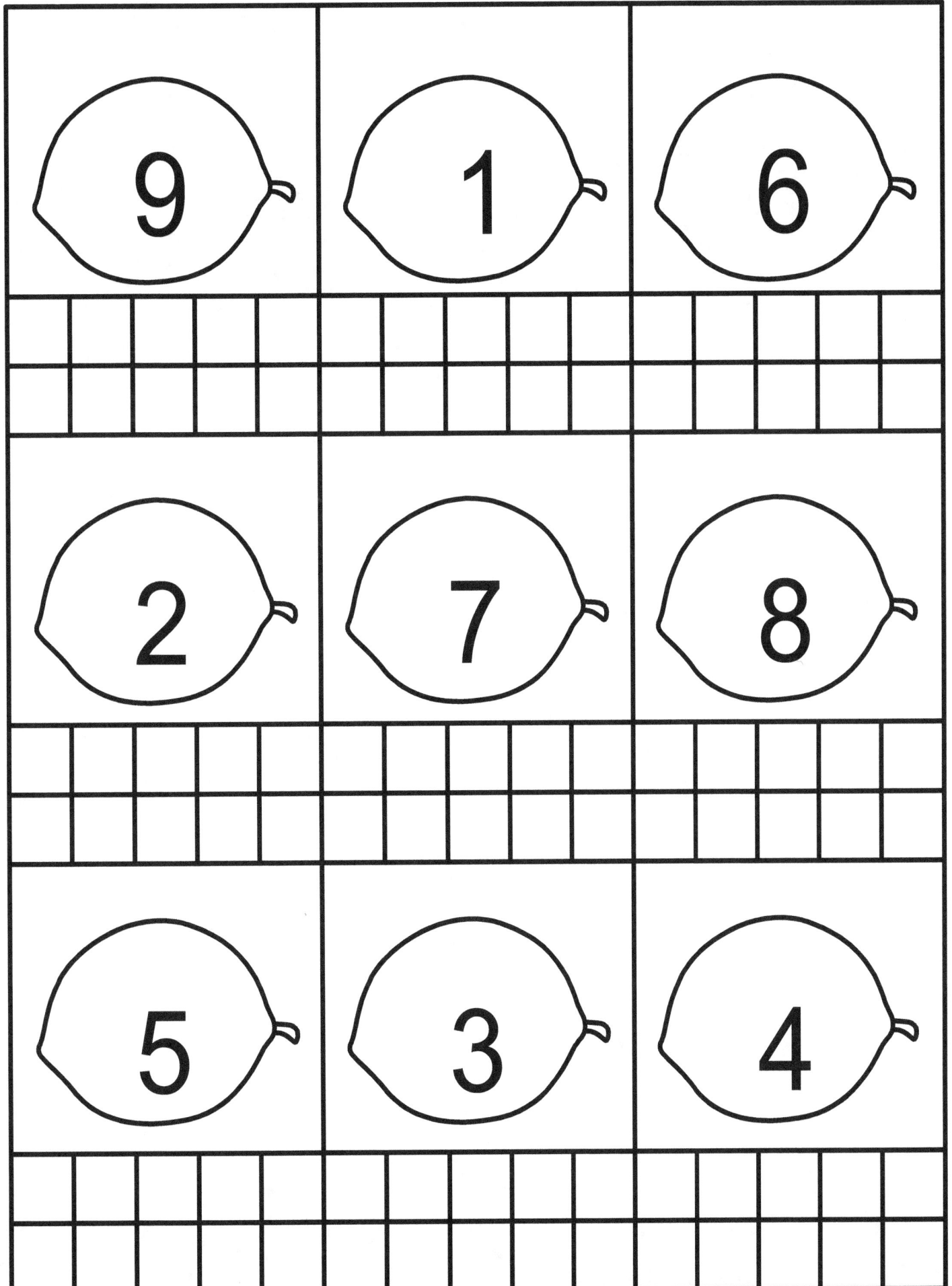

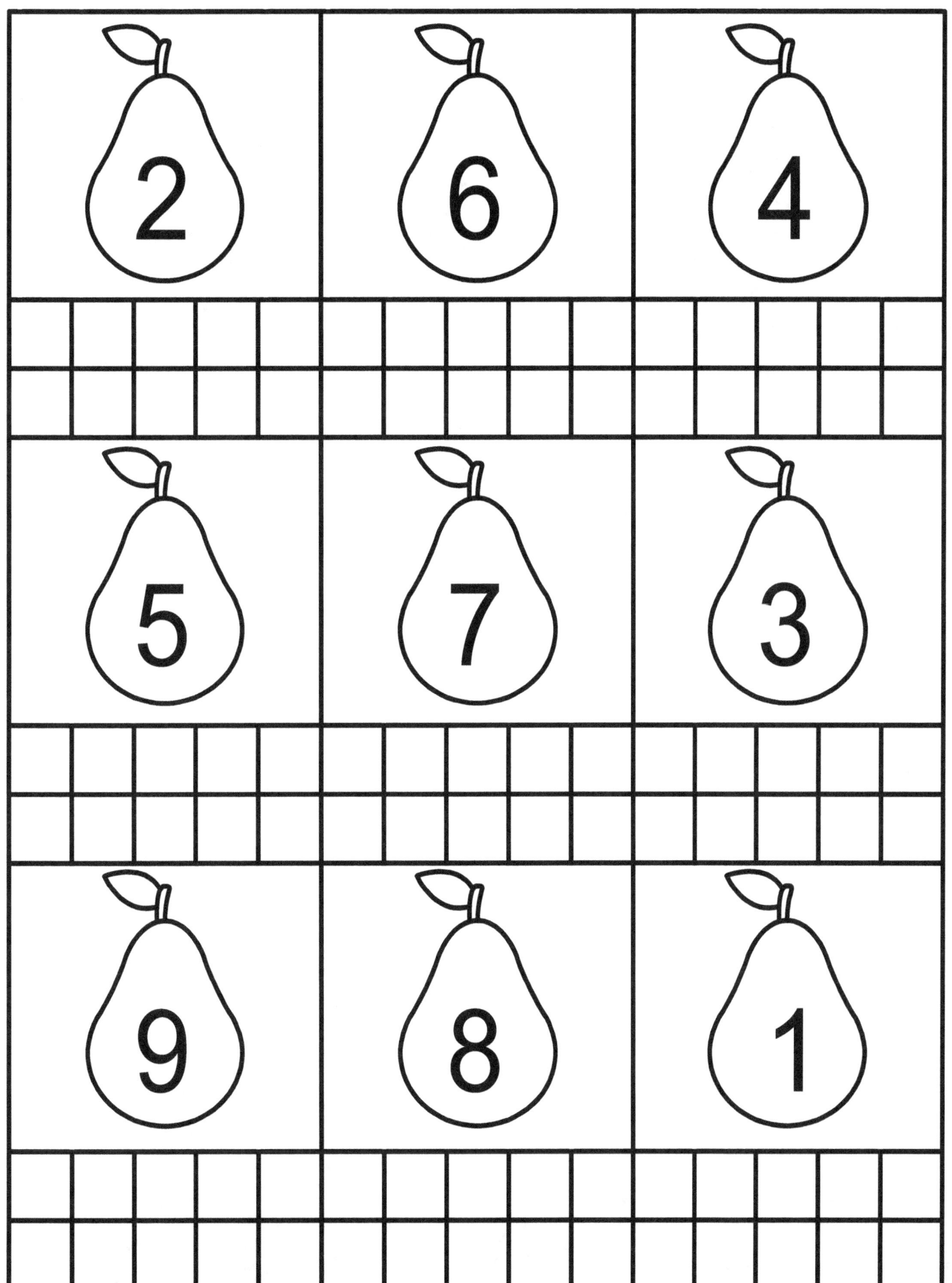

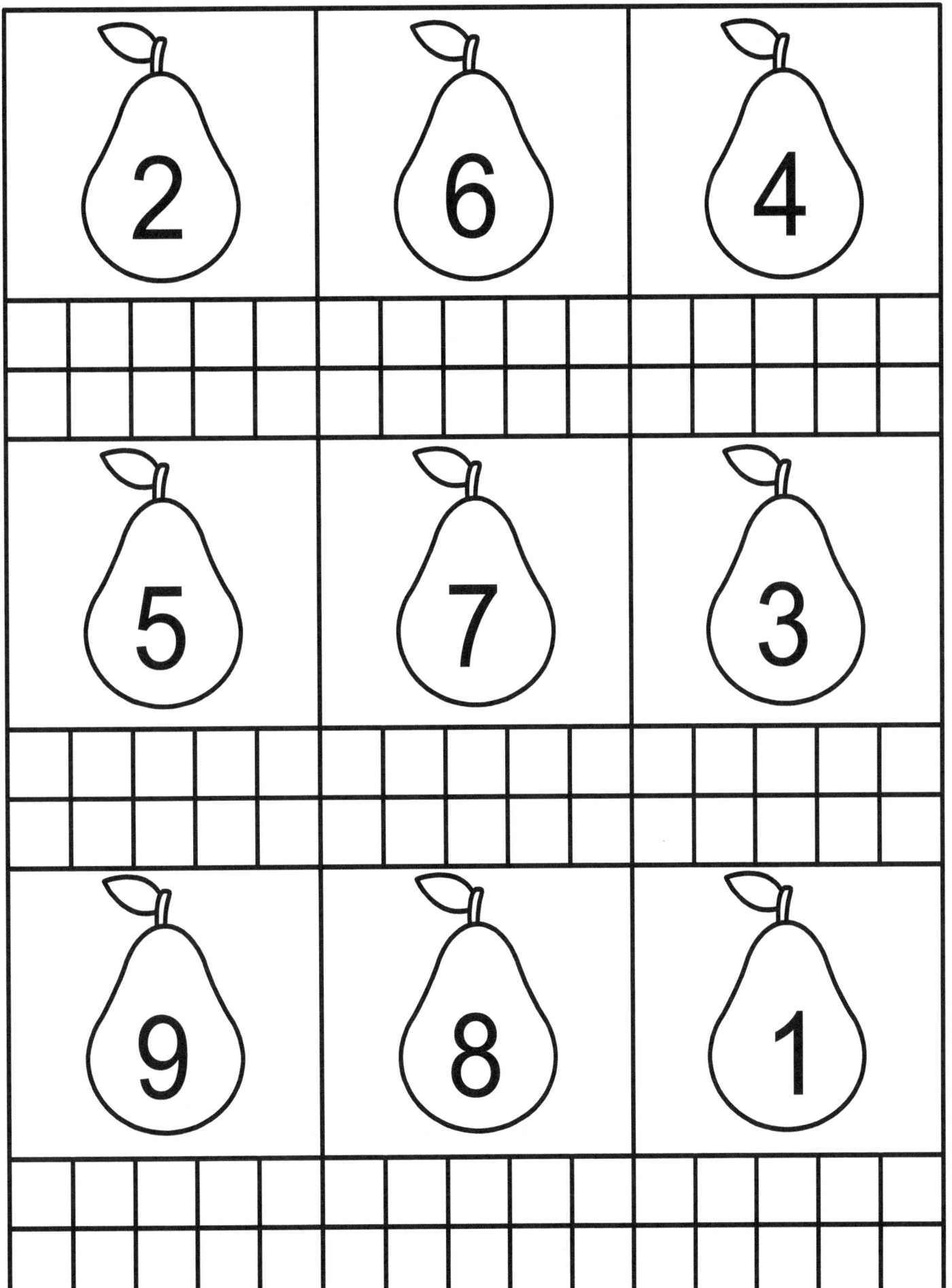

# Match the Number

Draw a line from the number to the matching set of objects.

2

4

3

5

8

# Counting

Count the objects. Write the correct number in the box. Color the objects for fun.

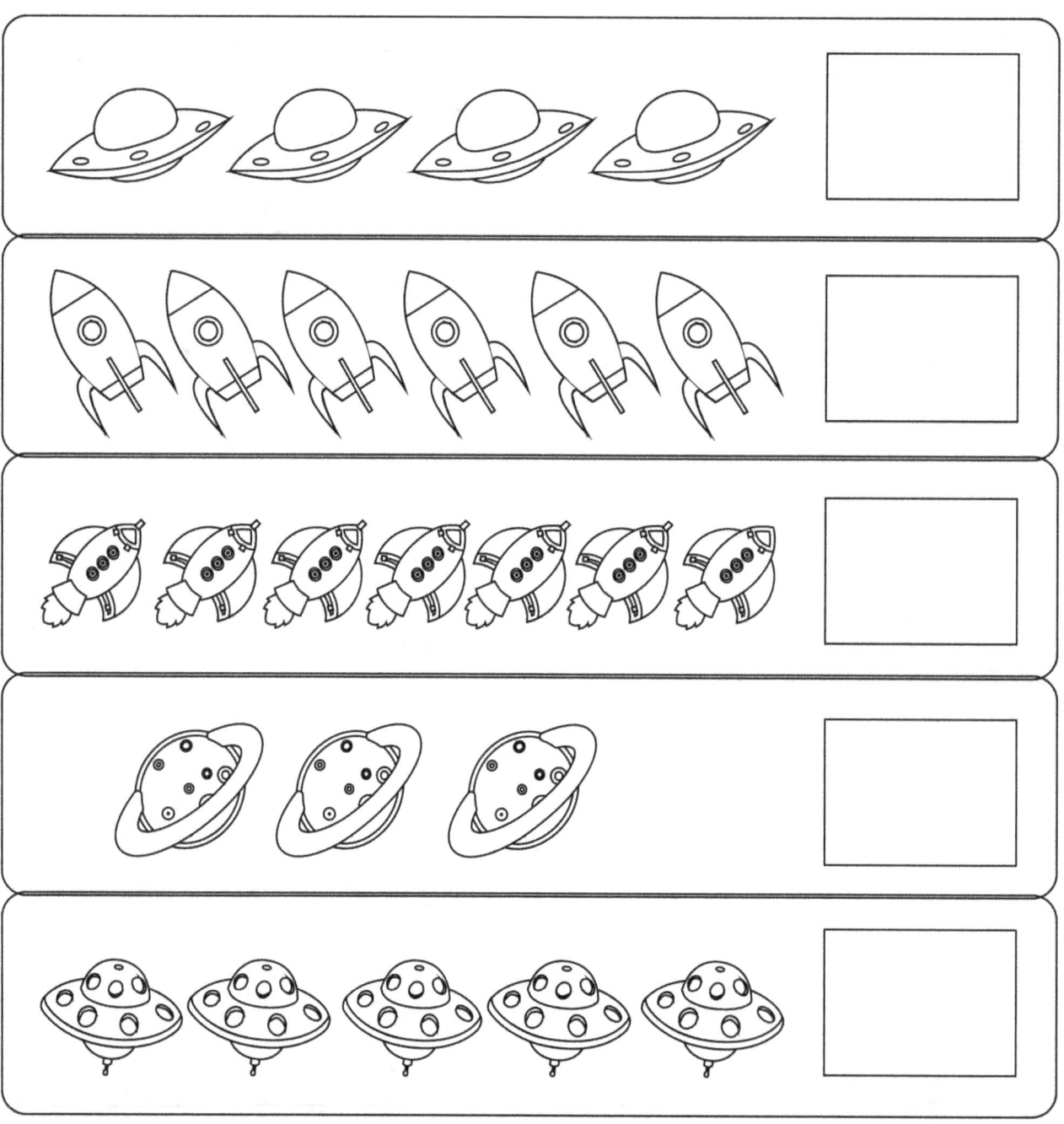

# Coloring Addition

Find the answer tio each problem and write in the given box. Use the numbers to color the pictures.

$6 + 3 =$ ☐   Yellow

$4 + 4 =$ ☐   Green

$2 + 2 =$ ☐   Red

$5 + 2 =$ ☐   Blue

$8 + 2 =$ ☐   Orange

$3 + 2 =$ ☐   Pink

# Subtract Numbers

Find the spaceship. Subtract the numbers in each box and color the spaceship with the correct answer.

6 - 3 =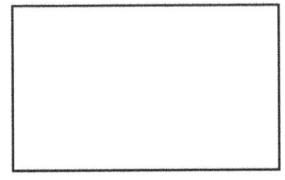

8 - 5 =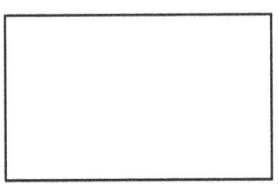

4 - 2 =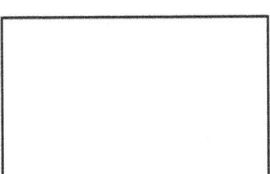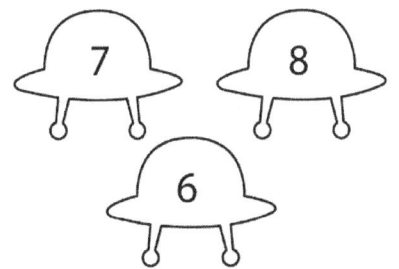

9 - 2 =

# ADDING

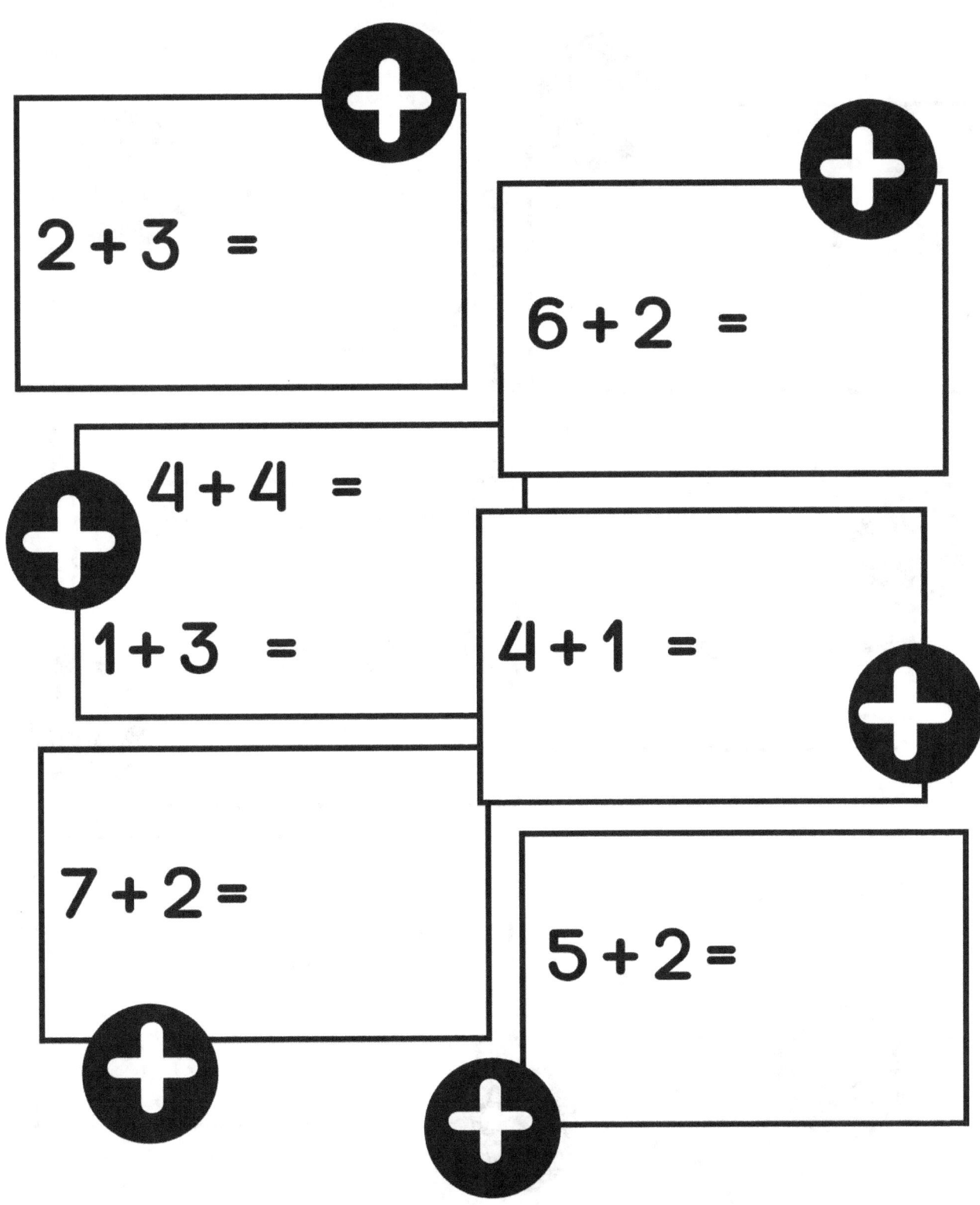

2 + 3 =

6 + 2 =

4 + 4 =

1 + 3 =

4 + 1 =

7 + 2 =

5 + 2 =

# SUBTRACTING

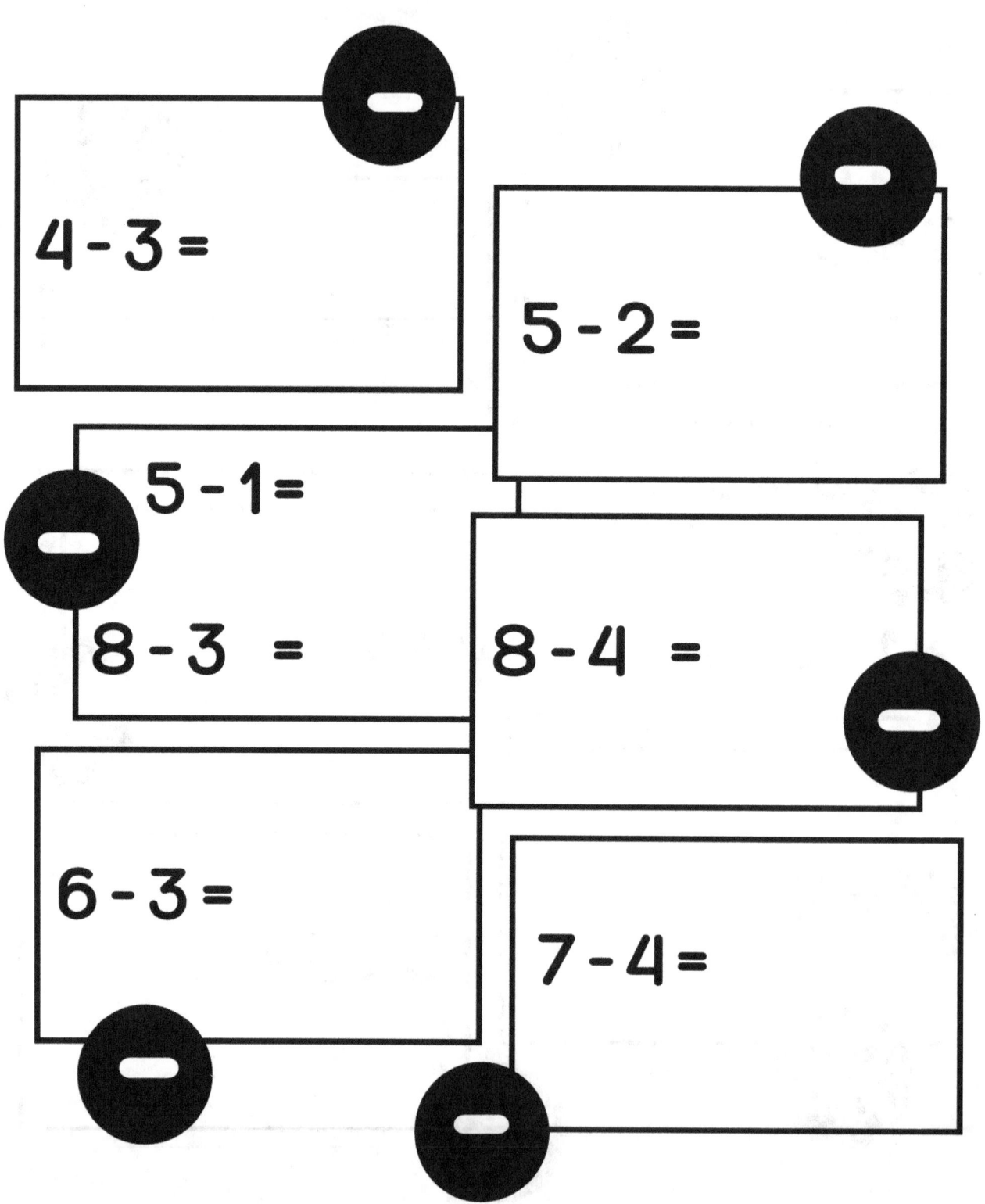

$4 - 3 =$

$5 - 2 =$

$5 - 1 =$

$8 - 3 =$

$8 - 4 =$

$6 - 3 =$

$7 - 4 =$